Project management for product and service improvement

Andrew Greasley
Technology Faculty, Open University

Butterworth-Heinemann
Linacre House, Jordan Hill, Oxford OX2 8DP
A division of Reed Educational and Professional Publishing Ltd

☘ A member of the Reed Elsevier plc group

OXFORD BOSTON JOHANNESBURG
MELBOURNE NEW DELHI SINGAPORE

First published 1997

© Reed Educational and Professional Publishing Ltd 1997

All rights reserved. No part of this publication
may be reproduced in any material form (including
photocopying or storing in any medium by electronic
means and whether or not transiently or incidentally
to some other use of this publication) without the
written permission of the copyright holder except in
accordance with the provisions of the Copyright,
Designs and Patents Act 1988 or under the terms of a
licence issued by the Copyright Licensing Agency Ltd,
90 Tottenham Court Road, London, England W1P 9HE.
Applications for the copyright holder's written permission
to reproduce any part of this publication should be
addressed to the publishers

British Library Cataloguing in Publication Data
Greasley, Andrew
    Project management for product and service improvement
    1. Industrial project management
    I. Title
    658.4'04

ISBN 0 7506 3769 2

Library of Congress Cataloguing in Publication Data
Greasley, Andrew.
    Project management for product and service improvement /
      Andrew Greasley.
    p. cm.
    Includes index.
    ISBN 0 7506 3769 2 (pbk.)
    1. Industrial project management.
    T56.8.G74   97-24160
    658.4'04–dc21   CIP

Typeset by Avocet Typeset, Brill, Aylesbury, Bucks
Printed and bound in Great Britain by Biddles Ltd,
Guildford and Kings Lynn

Project management for product and service improvement

Dedicated to Teamwork

Contents

Preface	ix
1 Value Analysis/Value Engineering	1
Aims	1
Introduction	1
A description of the techniques	1
What can be achieved	2
What cannot be expected	4
When can the technique be used?	5
Project initiation and selection	5
Value, function and cost	6
Value	6
Functional importance	7
Functional cost	9
A project selection method	10
Project selection in the manufacturing sector	10
Project selection in the service sector	13
Choosing between technically equivalent projects	15
Assessing the importance and cost of 'functions'	17
The importance of functions in products and services	17
Function costs in products and services	19
Generating new ideas in products and services	20
Presenting VA conclusions	27
Objectives	29
2 Financial appraisal methods for technically feasible projects	30
Aims	30
Introduction	30
The payback period	31
Average gross annual rate of return	32
Average net annual rate of return	33

	Discounted cash flow	34
	Compounding	34
	Discounting	35
	Net present value	35
	Applications of NPV techniques	37
	Objectives	39
3	**Using Network Analysis to implement projects**	40
	Aims	40
	Introduction	40
	Constructing networks	43
	Optimizing delays, working patterns and cash flows in projects	46
	Objectives	48
4	**Optimizing production rates and working practices**	49
	Aims	49
	Introduction	49
	Premature attrition of plant, equipment and tooling	49
	Using the approach in other situations	54
	Non-recovery of tooling plant and equipment costs	57
	Under-utilization of resources in the service sector	58
	Optimization of fixed and variable costs	59
	Objectives	65
5	**Displaying data and calculating averages**	66
	Aims	66
	Collecting and displaying data	66
	Tabulated data	67
	Pie charts	69
	Histograms and bar charts	72
	Graphical correlations	75
	Equations of lines on graphs	80
	Manipulating power law and exponential relationships	81
	Fitting equations to unusually shaped graphs	84
	Measures of central tendency	84
	Arithmetic mean	84
	Mid-range value	86
	Median	86
	Mode	87

Geometric mean	87
Grouped data	88
Objectives	91

6 Probability effects in projects 92
Aims 92
Probabilities of events 92
Acceptable variations in events 93
Unacceptable variations 94
What to do about variable quality 96
Combining probabilities 98
Risk analysis in safety critical components and systems 102
Risks in project implementation 104
Probability distributions 107
   Rectangular distributions 108
   Symmetrical binomial distributions 109
   General binomial distributions 110
   Probability effects in response rate calculations 110
   Probability effects in sales team planning 113
   Probability effects in mass production 115
Problems 117
Answers 118
Objectives 121

7 Standard deviations and normal distributions 123
Aims 123
Standard deviation 123
Means and standard deviations of probability distributions 127
   Alternative formulae for the mean and standard deviation
   of a probability distribution 128
From the symmetric binomial distribution to the
'normal' probability distribution 131
Using the area under the normal curve 136
Fluctuations in steady data and dealing with proportions 137
Null hypothesis 138
Levels of significance in surveys 140
Standard error of a mean 142
The significance of correlations between factors in projects 143
   Example 145
Defining the correlation and other statistical comparisons 146

Statistical process control applied to projects	147
Example	148
Objectives	153
Index	155

Preface

Current approaches to life in our industrialized society often involve the creation of a 'project' as a means to bring about an improvement. Such projects vary in size from, say, the production and certification of a new airliner down to a quick mail-shot about a cheap holiday offer.

However projects do not just happen. Successful ones are chosen carefully, analysed to maximize benefits, implemented logically and kept in control until their termination.

Projects are not always new ventures. Most are initiated to bring about relatively minor improvements to existing products and services. Unfortunately the 'divisions of labour' and specialist skills that are needed in large organizations, usually mean that no individual can accomplish all the project aims. With complex products and services, it is rare that even a small team has all the required knowledge, flair and determination to drive a project through to its best possible conclusion. Frequently there are also conflicting secondary interests within such teams. Hence all project teams can benefit from advice on how to avoid problems and how to organize the approach.

The availability of computer software means that it is often possible to process data and achieve results without a full understanding of the fundamentals behind the programme. This can often achieve the desirable effect but can also lead to mistakes when results are stretched beyond feasible limits. It may also create a feeling of insecurity and fear to question the system when apparently illogical developments occur.

Much of the content of this book is suitable for incorporation into PC software and some of the techniques are available as a matter of routine amongst commercial packages. However it is the purpose of this text to introduce several project oriented topics at a basic level upon which more complex procedures may be built.

The topics addressed are presented at undergraduate level and include many examples relevant to design and manufacture. Wherever possible, the principles have been extended for wider application. Hence much of this book is equally valid as a foundation for project work in the service sector.

The content has been influenced by involvement with several Open University Course Teams at undergraduate and postgraduate level. These too were project teams and acknowledgement is given here for the mutual benefits that evolved within them. Each chapter begins with a set of study aims and ends with a list of study objectives. Many worked examples are developed within the text. However the essence of project work is to adapt such procedures to suit the planned application. Hence the text is best studied in conjunction with a relevant examples class to demonstrate how the topics are adapted to suit local needs.

It is hoped that this text may spark an interest for further study in the areas of value engineering, financial appraisal, network analysis and production planning, project management, risk assessment and the statistical analysis of data.

Finally, acknowledgement is given for the word and diagram processing skills of the secretarial staff within the Technology Faculty of the Open University and to Tracy Bartlett and Cathy Playle in particular.

Andrew Greasley
Technology Faculty
Open University 1996

1 Value Analysis/Value Engineering

Aims

- To introduce the concept of 'value' in a product or service
- To introduce the concept of 'functions' within products and services
- To give a method by which projects can be selected
- To demonstrate how to conduct a Value Analysis on a Product, Process or Service

Introduction

A description of the techniques

Value Analysis and Value Engineering have their roots in Anglo-American manufacturing in the aftermath of World War II. Over the years **Value Engineering** has been the term used for **new products**, often produced on 'green field' sites where the emphasis is to get things right from the beginning. **Value Analysis** was the term used to describe the process of taking a careful look at **existing** products, processes and services to bring about improvements without radical changes. Both techniques are particularly important in large companies with wide product types and ranges, selling 'industry standards' that may have changed little since their launch. Occasionally the conclusion of a Value Analysis is that a total rethink is needed. The subject of the study may even be deemed totally obsolete and a full Value Engineering exercise on its replacement is then required.

In principle the techniques are equally valid when applied to the service sector. The provision of a service always has a value and an associated cost. Here the link between value and cost may be more difficult to quantify and may need much more detailed market investigation.

Examples from the service sector

The cost of a glossy brochure may add apparent value to a holiday package at the time of booking but if expectations are not realized then it may have been a wasted investment. In such circumstances the travel agent or tour operator must decide on the value they place on repeat business or if they are satisfied with 'once-off' custom.

The value of the 'back-up' service you offer after the sale of a component or product is particularly difficult to assess. At the extremes it is self evident. For example you don't mind buying another cheap calculator if one function button packs up but you expect to have a quick and reasonably priced repair to a personal computer with the same problem. In the middle ground assessing the business sense of the provision of product replacement, recall-repair and service calls is difficult and this is where these techniques provide assistance.

There have been variations in the terminology used. The approaches are included in 'Operations Research', 'Systems Analysis' and more recently in 'Quality Function Deployment'. However it is considered that the term 'value' is fundamental within the description and will be retained in this approach.

What can be achieved

Improved products and services
These are the obvious benefits. Critics of the technique often state that 'good engineers and managers are constantly performing VA assessments'. They may well be. Indeed, projects that are under careful, long-term scrutiny by the best informed team or individual are difficult to improve upon. However, 'cash cows', that bring in revenue that is taken for granted have a habit of becoming 'lame ducks' unless they receive regular and adequate attention.

The identification of new markets for existing products, improved value that eliminates competitors and new products for existing markets is unlikely to emerge from normal business operation without some effort. The ideas for such developments may occur by chance but these techniques attempt to eliminate the uncertainties and offer a methodology for logical and organized, product and service improvement.

Improved co-operation and communication
The necessity to operate in multi-discipline teams towards a common objective has been observed to have the spin-off of improving workforce morale, especially in large companies. There have been projects where very few changes have been made to products, processes and services. Improvements in co-operation have been all that is necessary. Hence it is best to use a 'neutral' leader of the VA team. This is essential when disputes between different areas and individuals are known to exist. Even when definite physical changes are needed in products and services it is best to choose a leader who will be able to see when personality and other indirect problems may jeopardize project outcomes.

Identification of poor control and information
During normal operation the VA team will identify the absence of critical information which is needed before improvements can be made. Such information deficits come in three broad categories:

(a) Limitations to the fundamental knowledge of the product or service. For example, inadequate stress analysis of a structural product or the unknown business gain of a 24-hour help-line.

(b) Quality and reliability data where the product or service is jeopardized by unknown quality issues. It is assumed that the known quality issues have been attended to by 'quality circles' or other working practices. However it is unlikely that an internal quality team would always be aware of changing market situations. For example the planned introduction of a new 'standard' at national or international level should be known of at the earliest opportunity. Searches for such information would be part of a full VA investigation.

(c) Quantifiable opinion on what the exact needs of the market are and how these needs may be fulfilled within the product or future service. This may or may not be achieved by a sales dominated information route. Sales personnel rarely say 'what is wrong with our product' to potential customers! Hence they may be unaware of the real weaknesses within the product or service.

It is the role of the VA team to initiate the work required to improve this information and control base. However this too has a cost associated with it and if the level of research and development needed is too high or the quality issues too demanding in terms of return on investment, then the VA team would recommend abandoning the project. It may recommend leaving the market sector or even the business. Information is often considered as a useful resource. However if it costs you so much to obtain it that you will not get your money back by using it, it is comparable to treasure buried so deep that it is not worth excavating.

Creation of entrepreneurial conditions
The initial entry into a market or radical change within it, is often guided by individuals or small teams who have most of the current information and experience at their fingertips.

In large organizations such teams often get dispersed and individuals move on. However by establishing a small multi-discipline VA team it is often possible to re-create these conditions even though the initial enthusiasts are long gone and the product or service has been taken for granted for some time. Many large organizations say that identification of a 'common aim' generates its own corporate motivation. Occasionally the motivation and atmosphere created gives more widespread benefits.

What cannot be expected?

The team itself cannot create its own fundamental skills and experience. If expert advice is required then an expert must be recruited to the team or a sub-project commissioned to supply the fundamental information. Alternatively a team member may be delegated to train-up on a topic. However this will clearly delay the project.

Critics often pour scorn on to the mathematical rigour behind the technique. The mathematics behind the concepts used are indeed very simple and often based on subjective opinion. In defence, when proponents ask for suggestions of alternative approaches few are offered.

When can the technique be used?

The technique can be used at the conceptual design stage, the detail design stage or it can be applied to mature products and services.

When it is applied to existing products and services the outcome will usually be a change to the details of the artefact or offering. In certain circumstances the outcome may require a change to the conceptual design. This may occur when a technology has already been pushed to its limits or a service has been refined to such a level that changes may bring such small benefits that they cannot be justified.

VA teams have their most difficult decisions to make in such circumstances. Debates on whether to continue with proved technology and systems or not, are common. It is frequently hoped that detail changes can bring acceptable improvements since the alternative is to invest in a new concept and discard much of the corporate experience.

Occasionally external influences make the decisions easy. For example the refrigeration and air conditioning industry has used improved detail design to counter political and social pressure over the use of chlorinated-hydrocarbons. A total ban on their manufacture would require a conceptual change in order to introduce a new 'cooling' concept. The aerosol industry has had to be more drastic. For certain products the concept hasn't changed. Minor changes have allowed more acceptable propellant gases to be used. However for other products other methods of atomization have been introduced as conceptual design changes.

'Officianados' of design methodology often get into long debates about where the border between conceptual design and detailed design is placed. Do not let such debates slow down your team!

Project initiation and selection

Rarely will there be one obvious candidate for a project. For projects within an established manufacturing or service base there will usually be several products or services that need attention. In a 'green-field' situation there will usually be many detailed solutions proposed within each conceptual design which may require different capital equipment and personnel. Hence the flexibility of the equipment and people, which enables the unit to function in different areas, may be an important consideration by a VA team concerned with long term market planning.

Making selections at the conceptual design stage requires even more care. Choosing between different conceptual solutions often means committing large investments over long periods to develop a particular approach. Hence confidence is needed that your concept has a proved long term advantage over alternative approaches. For instance the recent massive long term investments in satellite communication are based on the premise that existing surface cable technology is at its limit of capability and the concept of sending communications through the earth's core will not be viable for a considerable time, if ever!

Table 1.1 (p. 11) provides a formal approach that enables a choice between projects to be made. Where this has already been done the team must consider the central issues in the project which we now consider.

Value, function and cost

Value

The technique hinges round the balance of three items in the title. They are usually compared using this relationship:

$$\text{Value} = \frac{\text{Functional importance}}{\text{Functional cost}}$$

The units of measurement in this relationship are difficult to define precisely. However that does not matter as long as like is compared with like and the effects of one do not numerically swamp the effects of another.

The value of a product or service always contains a proportion of the value types shown below:

Value Analysis/Value Engineering 7

(1) **Service or use value**, the simple ability of the device or service to perform the desired task.
(2) **Prestige value**, consists of properties that make ownership of the product or receipt of the service preferable.

The relative amounts of these values depends on the precise nature of the product or service. Certain items of a strict industrial nature have negligible prestige value. A 10 MW turbo generator is just that, nothing more, nothing less. Long term reliability and good maintenance performance may influence choice between alternative suppliers. However this would still be strictly a 'service-value' decision.

Conversely certain products are associated with high prestige value. Sports equipment is notorious in this respect. Baseline performances (if you'll pardon the pun!) are often equivalent. However certain products gain prestige due to endorsements and advertising campaigns.

In the service sector similar distinctions exist. For instance, long-haul aircraft operators can provide high service value by putting seats close together. However they risk reducing prestige value in both real and imaginary terms if passengers arrive in a poor physical state or frame-of-mind. Hence the attention to 'in flight service' as the operators try to increase perceived prestige value with a minimum drop in service value.

Service value is usually improved by reducing the cost of the function. However it can also be improved by offering better functional importance for similar functional cost. There is overlap between service and prestige value. For example the latter can be improved by over-specifying the initial service performance to give exceptional long term reliability.

Functional importance

Before this can be estimated it is important to analyse functions in more detail. Whatever the function, it is part of a general **function tree** which is best described by means of example as shown in Fig. 1.1. Here the product of a **mousetrap** and the service of **pest control** are shown side by side. You move up the trees by asking the question 'Why?' and down by asking 'How?'.

8 Value Analysis/Value Engineering

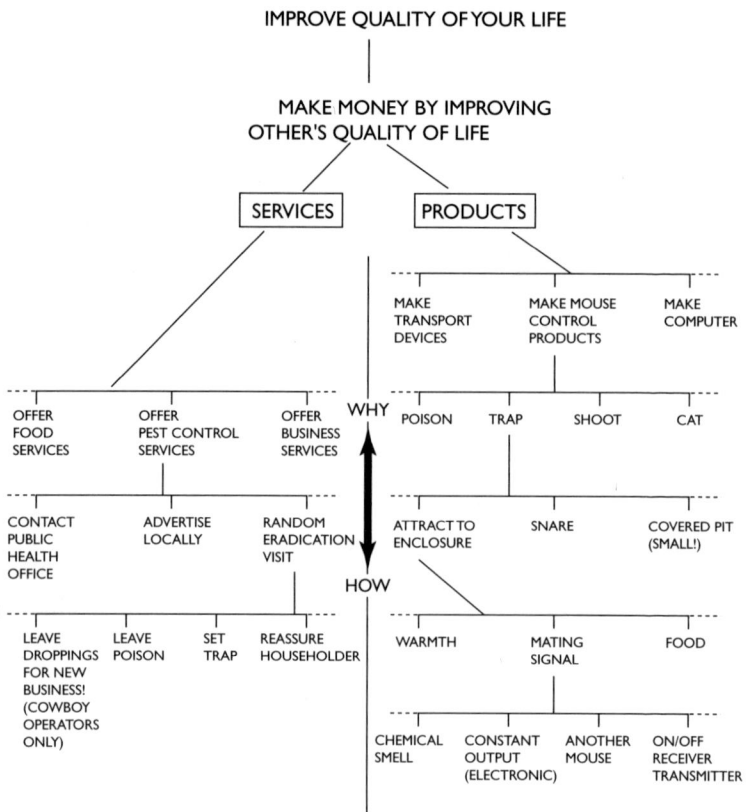

Figure 1.1 Part of a general function tree selected for a mousetrap producer and a pest control service

Once the need for providing a particular overall function has been justified and placed in the correct level on a function tree then the means of providing that function can be investigated further.

Such analysis is not as easy as it first appears because many components in products and services contribute to several functions. Some of these may be **primary functions** without which the product or service would not work at all, **secondary functions** which assist the primary function and **aesthetic functions** which contribute to the prestige value of the product or service.

For instance the steel of a car's body work has a primary function which is structural. A grey paint finish is provided to contribute the secondary function of corrosion resistance. This gives the primary function longevity. Finally the glossy, coloured, top coat provides the aesthetic function. In the service sector a bank cashier's primary function is to provide cash transfers. The redirection of customer enquiries about other financial matters is a secondary function and the ability to conduct these affairs in a pleasant manner is an aesthetic function that contributes to the prestige value of the particular bank.

In the sections that follow, semi-quantitative techniques will be presented that enable these functions to be given an importance ranking. This can be turned into value once the cost of the function can be established.

Functional cost

This is simply the monetary cost of providing a particular function in a product or service. At this stage do not be concerned that the concept of 'function' has not yet been fully developed beyond the concept of the function tree and the different types of function. This will be done soon enough. Such costings may be clouded by local accountancy issues and part of the analysis should be to examine if these are crucial to the outcome. For example it is very common to split costs into 'fixed costs' and 'variable costs'. The fixed costs cover the aspects of providing the basic factory or office and do not change much with the level of production or business. The variable costs change with the level of activity and cover things like materials, tools, the wages of the direct workforce, etc. The fixed costs are the most difficult to apportion between different products and services. Hence being able to associate them to particular functions **within** a product or service is often very imprecise. Several costing methods are given in Chapter 4 for whole project analysis. However, it is best to stick with the local accountancy system for detailed calculations and try to apportion a share of the fixed and variable costs to particular functions. For comparisons of projects within an establishment, this is usually adequate. Nevertheless, caution is needed and the following example highlights the sort of dilemma that can occur. A small gold foundry area exists within a much larger non-ferrous metal foundry. Do you apportion the fixed costs using the floor space (a common method) or by the turnover (an equally common method)? Clearly the floor area would underestimate the costs because the small precious metal area would benefit from the presence of the large factory (security, fuel bills, indirect workers). The turnover would grossly overestimate the costs to be attributed to the gold casting area, because the relatively high intrinsic costs involved would attract an unfair amount of the service costs. This would not matter for projects within the gold foundry area but it would influence the outcome if comparisons were made on the return on investment across the whole product range; or worse still if the gold foundry associated with the large factory was compared with a small self-contained gold foundry that was in competition with it.

In the service sector very similar problems may arise when private health care is conducted within a small area of a large hospital linked to a national system. Extracting an accurate cost of such provision is very complex because of the interactions between the two systems. Placing a value on the symbiotic nature of the association is also difficult.

If the true cost of a component or service can be established then the problem of which function it contributes to remains. For instance the primary function of a vehicle dashboard is to support instruments. There are safety aspects that need consideration as a secondary function. The clarity of the instruments and crash behaviour need to be considered. However, dashboards are much more complex than they need to be because they have a significant aesthetic function in their role as interior trim parts. The worked example given in pp. 17–19 will demonstrate a method producing a functional cost breakdown for such components and services.

A project selection method

Project selection in the manufacturing sector

This method is best explained by way of example. Table 1.1 gives a blank selection chart where points are awarded for various aspects of each project. The aspects are explained below for projects involving manufactured products.

Table 1.1 Project selection chart

Selection aspects	Points given	Project										
		A	PTS	B	PTS	C	PTS	D	PTS	E	PTS	F	PTS
Annual sales (£) x design life expectancy	1→5												
Competition* low	1												
med.	2												
high	3												
Time to study 6 months	1												
3 months	2												
1 month	3												
Actual profit margin / Potential profit margin	1→5												
Individual product	1												
Limited range	2												
Full ragne	3												
State-of-art high	1												
med.	2												
low	3												
X factor × sale × life	1→5												
Prestige potential low	1												
med.	2												
high	3												
TOTAL POINTS													

Selected projects		Reasons for selection

*Points are awarded here for projects involving existing markets. For new ventures into new markets the points should be reversed.

Annual sales income × Design Life Expectancy (1–5 points)

This gives the total revenue potential for a particular product. It is not worthwhile working on products with short design lives and low incomes. Hence for a selection from five projects these should be ranked 1–5 with the product with highest revenue potential getting 5 points. For a selection from greater numbers of projects they should be banded together in 5 revenue bands and each member given the points value attributed to that band (again 1-5).

Competition (1–3 points)

If you are already in the market it is best to work on projects that are in competitive sectors because it is pointless spending money improving products that have little competition and sell in the market place anyway. If you are not in the market then you may wish to reverse the points allocated so that projects that get you established firmly in open markets with little competition receive the best marks.

Time to study (1–3 points)
Short development times are preferred for projects. If a preliminary skirmish reveals that new tooling is required (e.g. injection moulds) then this means the projects will be much longer in bringing in the revenue. Hence the implications of the project need to be identified as early as possible. However many businesses have been lost completely by putting off investment in new technology. For example, injection moulded dinghy and canoe hulls changed the market almost overnight once the moulds had been perfected. This was to the detriment of glass fibre specialists who had persevered with hand-lay-up methods and were unwilling to risk capital on a new technology to the boat building industry.

Profit margins (1–5 points. Once again surveys with more than 5 projects should employ 5 categories)
Projects are preferred where the potential profit has not been achieved and the ratio given is low. For new products 'Actual profit margin' should be replaced by 'Likely profit margin' and 'Potential profit margin' should be replaced by 'Maximum possible profit margin'. Poor performance can often be attributed to 'over engineered' products or too many intermediates in the sales and distribution system. Cash flow problems may also contribute to the situation. Many potentially sound projects fail when the setting up of a large sales and distribution network has been done without taking 'bad debt' levels and late payments into consideration.

Product range (1–3 points)
Projects that involve products that may produce spin-offs in other parts of the product range are favoured.

State-of-art (1–3 points)
This refers to the level of development or maturity of the product. Products that have been worked on recently are unlikely to yield much improvement and gain low points.

'X' Factor multiplied by revenue potential (annual sales × life)
(1–5 points, use 5 categories for more than 5 projects)
This estimates the total savings, or improved value that should result from the project. Although the 'X' factor should be decided by the project team, the following table gives some guidance. This balances the benefits of large sales revenue against the improvements that might be expected. Large values therefore gain the most points since minor improvements on major sales lines may still be very worthwhile.

Project	'X' Factor
Fresh investigations on 'Low state-of-art' products	20% or less
Repeat investigations on 'Medium state-of-art' products	10% or less
Repeats on mature products	5% or less

Prestige Potential (1–3 points)
These points should be allocated to strengthen projects that are going to build on the prestige already developed in other areas. For instance if you already make good teapots you could capitalize on this by introducing coffee jugs, even though the competition and their 'state-of-art' are both high. Simply adding your name to another type of product may bring rewards! However make sure it's a good value coffee pot or your existing market reputation may suffer.

As usual these techniques are best demonstrated by example and Table 1.2 gives some data for the product range for a company in the electrical product industry. Table 1.3 gives the project selection outcome.

Project selection in the service sector

The interpretation of Table 1.1 for the service sector requires careful consideration. For example the 'design life of a product' converts to 'how long will this particular level of service be suitable?' Most of the other categories in Table 1.1 have obvious comparisons in the service sector. Table 1.4 shows how the scheme

	Annual sales k£	Design life years	Revenue potential k£ yr	Competition level	Typical study time	Actual profit / potential profit	One off product or part of range	Current state of development	Savings estimate level X%	Prestige potential
A Travel hairdryer	900	3	2700	v. high	3 months	$\frac{15\%}{20\%} = 0.75$	One-off	v. high	2%	None
B Foodmixer	1000	5	5000	high	6 months	$\frac{25\%}{30\%} = 0.83$	Med. range	medium	10%	High
C Drill motor part assembly	500	7	3500	high	3 months	$\frac{10\%}{10\%} = 1.0$	Fits a large range	high	5%	None
D Electric razor	2000	10	20 000	medium	3 months	$\frac{20\%}{30\%} = 0.66$	One-off	high	5%	High
E Electric clothes brush	10	12	120	low	1 month	$\frac{15\%}{20\%} = 0.75$	One-off	low	15%	Medium

Table 1.2

could be re-formatted for a traditional restaurant planning to either expand, or introduce fast food or introduce haute cuisine or introduce a totally new approach.

Table 1.3

Selection aspects	Points given	Project A H. Dryer	PTS	B Mixer	PTS	C Drill	PTS	D Razor	PTS	E Brush	PTS
Annual sales (£k) × design life expectancy	1→5	2700	2	5000	4	3500	3	20 000	5	120	1
Competition low / med. / high	1 / 2 / 3	v. high	3	high	2	high	2	med.	1	low	0
Time to study 6 months / 3 months / 1 month	1 / 2 / 3	3 m	2	6 m	1	3 m	2	3 m	2	1 m	3
Actual profit margin / Potential profit margin	1→5	0.75	4	0.83	2	1.0	1	0.66	5	0.75	4
Individual product (Ind.) / Limited range (L.R.) / Full range (F.R.)	1 / 2 / 3	Ind.	1	L.R.	2	F.R.	3	Ind.	1	Ind.	1
State of art high / med. / low	1 / 2 / 3	v high	1	med.	2	high	1	high	1	low	3
X factor × annual sales × life	1→5	54	2	500	4	175	3	1000	5	18	1
Prestige potential low / med. / high	1 / 2 / 3	none	0	high	3	none	0	high	3	med.	2
TOTAL POINTS			15		20		15		23		15

Selected projects	Reasons for selection
D Electric razor = 23 pts	Top score, plus large sales revenue, basic design life is long, med.-short project needed

Choosing between technically equivalent projects

Scores within a few points of each other on such schematic appraisals are likely to be technically equivalent. In such cases the project should be selected using a cash flow technique that will optimize the rate of return of the capital outlay associated with the project. Such techniques are explained in detail in Chapter 2.

16 Value Analysis/Value Engineering

Table 1.4

Product format	Points given	Restaurant development format
Annual sales (£) × design life expectancy	1→5	Sales estimate × number of years the food will be in vogue
Competition low	1	High – many similar outlets locally.
Competition med.	2	Med – few similar outlets locally.
Competition high	3	Low – no similar outlets locally.
Time to study 6 months	1	
Time to study 3 months	2	Time to establish new clientele.
Time to study 1 month	3	
Actual profit margin / Potential profit margin	1→5	New venture so look for projects that are likely to achieve high values of this ratio since potential profit levels are uniform in the food sector.
Individual product	1	Possibility to link with other
Limited range	2	restaurants or chains via franchise
Full range	3	agreements?
State of art high	1	Is the local population ready for this type of food or development that is not being provided? If so the state-of-art low.
State of art med.	3	
State of art low	3	
X factor × annual sales × life	1→5	Your project is: single outlet X = 20% few more outlets X = 10% many outlets X = 5%
Prestige potential low	1	Will the project build on the current reputation (3 points)? No change (2 points) May reduce image (1 point)
Prestige potential med.	2	
Prestige potential high	3	

This has been a general approach about the benefits that can be expected in different types of project and how differing aspects need consideration in order to develop an overall semi-quantitative ranking. The chapter continues with a methodology by which the anticipated changes can be achieved. The techniques are best displayed by example and may be applied once a particular project has been selected on technical and financial grounds.

Assessing the importance and cost of 'functions'

The importance of functions in products and services

Once a project has been selected the VA team should not look at any manufactured product as an assembly of component parts. Services should not be considered as a staff and facilities list. Instead the topic should be considered as a list of functions. These may be the primary, secondary or aesthetic functions previously considered.

Consider the domestic 13 amp electric plug shown in exploded form in Fig. 1.4. It is typical of those that have been fitted by certain product manufacturers (when they choose to!) and may also be obtained separately. The component actually provides the following functions:

A   Temporarily connect an appliance to an outlet socket   (primary)
B   Semi-Permanently connect appliance to supply lead   (primary)
C   Provide electrical integrity (long term)   (primary)
D   Provide mechanical integrity (long term)   (secondary)
E   Disconnect supply on overload (fuse)   (secondary)
F   Be transferable for reuse on other appliances or
     changeable for differing outlet socket configurations
                                                            (secondary)
G   Look attractive/appropriate   (aesthetic)

With a small list it is sometimes straightforward to assess the relative importance of each function. However a method of comparing pairs of functions in what is sometimes called a 'triangulation method' often gives useful further insight. This is particularly useful in complex consumer durables and complex service provisions. For instance a video recorder has many secondary functions other than recording a TV programme and hotels provide much more than sleep and food. But what are the relative importances of all the other functions?

In such cases each function is compared with each other using the layout in Fig. 1.2. There is no upper limit for this method.

18 Value Analysis/Value Engineering

```
A )   A )   A )   A )   A )   A )
B )   C )   D )   E )   F )   G )

B )   B )   B )   B )   B )
C )   D )   E )   F )   G )

C )   C )   C )   C )
D )   E )   F )   G )
```

AND SO ON
↓

Figure 1.2   Triangular method of 'pair comparison' of functions

When the triangular layout has been completed the most important function in each pair is asterisked. The number of asterisks associated with each function is added up and converted to a percentage.

For the electrical plug functions the triangulation method gives the outcome shown in Fig. 1.3.

```
A*)  A*)  A*)  A*)  A*)  A*)
B )  C )  D )  E    F )  G )

B )  B*)  B*)  B*)  B*)
C*)  D )  E )  F )  G )

C*)  C*)  C*)  C*)
D )  E )  F )  G )

D*)  D*)  D*)
E )  F )  G )

E*)  E )
F )  G*)
F )
G*)
```

Function	Rank total asterisks	% importance
A	6	28
B	4	18
C	5	23
D	3	16
E	1	4
F	0	1(token value)
G	2	9
	21	≈100

Figure 1.3   Pair comparison for a 13 amp plug

The original function list was supplied in rank order by an electrical installations fitter and the triangulation method was done with marketing and product liability in mind. In the UK the two British Standards covering plugs and fuses should also be consulted in detail throughout such a project.

Notice how electrical integrity (C) has now been elevated to second place. Notice too that product styling (G) is considered more important than being able to transfer the plug to another appliance (F). Do you always take plugs off broken appliances? Have you ever discarded a plug because it looked old-fashioned? Indeed many manufacturers now mould plugs permanently onto equipment leads, making re-use impossible. This also guards against poor and unsafe installation procedures.

Function costs in products and services

Each component part of a product or service can contribute to several functions within that product or service and it is usually possible to break down the total cost of each component into portions that are each identifiable with the different functions to which the component contributes. Similarly, services can be broken down into portions whose costs contribute to different functions.

For example the costs of a uniformed doorman at a hotel could be split into essential help (primary function) security provision (secondary function), and prestige (aesthetic function).

In this way a table of function costs can be established from a list of component prices or service costs and a knowledge of how to split them. It is the latter part that requires some careful consideration within the VA team. The partial cost of a function can sometimes be established by asking the question 'What could the component or service cost if it did not have to provide this function?' Clearly the difference between this cost and the actual full cost relates to the cost of that particular function.

Such a breakdown of costs for the 13 amp electrical plug shown in Fig. 1.4 are given in Table 1.5.

The costs and importances calculated can now be compared and the bar chart layout shown in Fig. 1.5 is a convenient method of displaying the findings. The lengths of the bars are related to the cost and the relative importance of each **function**.

Hence this is much more informative than a simple components price listing.

It can be seen that transferability has a cost for negligible functional importance. Mechanical integrity is also slightly more expensive in relative terms than electrical integrity. The 'fuse' func-

20 Value Analysis/Value Engineering

tion is also very expensive compared to its importance.

Such histograms highlight the areas where benefits might accrue. But how do we bring about changes? First we generate some ideas.

Generating new ideas in product and services

Coming up with new design ideas is not always easy, especially on a mature product like an electrical plug that has evolved over the last century since the first generation of electrical power. Changes in national standards can necessitate essential changes, such as when the UK changed from round connection pins to rectangular pins. However for established products and services it is often quite difficult to obtain improvements. The generation of new ideas is usually the problematic stage. Hence there are a few ways of stimulating new ideas and evaluating them.

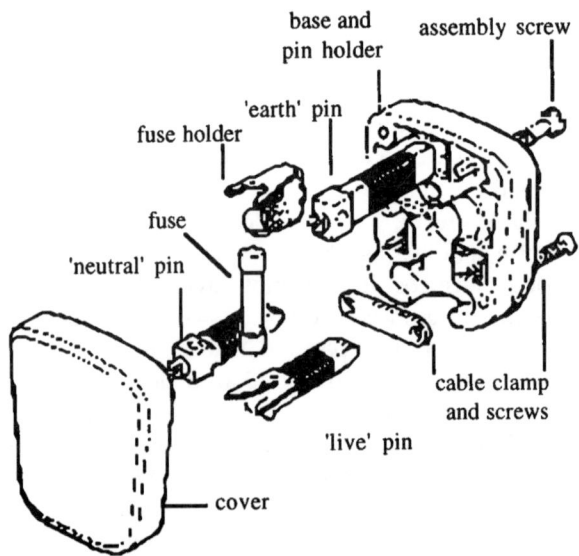

Figure 1.4 Exploded view of 13 amp electrical plug

Value Analysis/Value Engineering

Table 1.5 Function costs for a 13 amp electrical plug

	\multicolumn{7}{c}{Function}
	Outlet connection	Appliance connectivity	Elec. integ.	Mech. integ.	Fuse function	Transfer- ability	Appearance	Total component cost
	A	B	C	D	E	F	G	
Main base and pin holder	1.0	–	1.0	7.0	–	0.5	0.5	10p
Cover	–	–	3.0	0.5	–	–	0.5	4p
Cable clamp and screws	–	2.5	1	–	–	0.5	–	4p
Earth pin + screw	2.5	1.0	0.5	1.0	–	–	–	5p
Neutral pin + screw	1.5	1.0	0.5	1.0	–	–	–	4.0p
Live pin	1.0	1.0	0.5	1.0	0.5	–	–	4.0p
Fuse	–	–	–	–	2.0	–	–	2p
Fuse holder + screw	–	–	2.0	–	3.0	–	–	5p
Cover screw	–	–	1.5	0.5	–	–	–	2p
							Total product =	40p
Function total	6p	5.5p	10p	11p	5.5p	1.0p	1.0p	
Function cost %	15%	14%	25%	27%	14%	2.5%	2.5%	

Idea Stimulator Prompt List

Table 1.6 gives an example of a checklist that is relevant to the manufacture of small electro-mechanical devices. Clearly such a checklist should be made as specific as possible to the market sector in which it is to be used. Once again this technique is equally applicable to the service sector.

Table 1.7 shows the checklist modified for use in the service sector. It has been left as general as possible and can easily be adapted to suit the exact market sector.

The main output of such lists is not just new ideas. They also identify the need to gather and analyse particular information. This may come from several areas. For example a typical service report summary for a 13 amp electrical plug created by questioning several electrical fitters, is given in Table 1.8. Such reports should be used by the VA team to develop ideas and recommendations for improved designs. Careful scrutiny of the service comments in Table 1.8 will reveal several product improvement ideas.

22 Value Analysis/Value Engineering

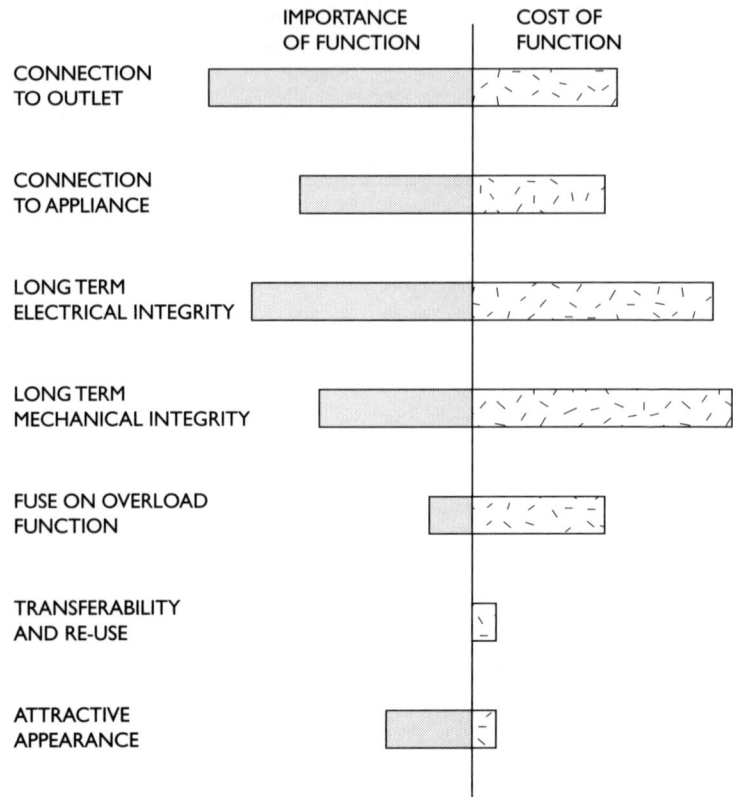

Figure 1.5 Function cost importance histogram for an electrical plug

Ideas Evaluation Action Chart
During the course of a project when ideas come flooding in it is good practice to formalize them using the chart given in Table 1.9. This one was taken from an electrical plug project.

Exposing where the costs lie in Manufactured and Assembled Products
Alongside a 'function analysis' a second approach can be used to highlight high cost aspects associated with a product.

Table 1.6 Ideas Stimulation Prompt List

Subject	Basic question	Subsidiary questions
Function	What functions are performed?	1. Can any function be eliminated? 2. Can any other way be found to achieve the same function? 3. Can another component incorporate this function?
Material specification	What is the material specification?	1. Can any other material be used? 2. Can any other spec. of same material be used? 3. Can a mixture of materials or surface treatment be used?

Material content	What are the dimensions of the materials used in the components	1. Can any dimensions be reduced?
		2. Is the part oversize?
		  (a) By calculation
		  (b) By comparison with associated parts
		  (c) By physical test
		  (d) By comparison with competitors' product
		3. Optimization problem – can any dimensions be increased and less costly materials used or vice versa?

Material value and waste	What percentage of basic material is wasted?	1. Can waste be reduced by making the blank nearer the finished size? 2. Can waste be reduced by minor design specification changes? 3. Can waste be reduced by changing the method of manufacture?
Dimensional limits	Which limits are critical?	1. Can any limits be relaxed to ease manufacture? 2. Can any limits be relaxed to allow alternative methods of manufacture?
Surface finish	Which surfaces are critical	1. Which are 'service' and which are 'aesthetic'? 2. Can 'textures' be used to cover over problematic surfaces?
Standards	Which standards are relevant	1. Are any likely to be slackened? 2. Are new ones about to be introduced? 3. Can we get overly tight standards reduced? 4. What benefits are gained from manufacture to the standards used?
Service history	What past problems have been reported?	1. Have competitors solved problems? 2. Will new designs eliminate problems? 3. Do we expect new problems with the new design?
Sales forecasts/reports	What aspects of the product limit sales?	1. Have any competitors successfully addressed this issue? 2. How saturated is the market? 3. Would any new functions in the product expand market share?

Often the costs of a product or service are concentrated within a particular area. These costs are not always associated with expensive parts but may be attributed to assembly, packing, finishing and other indirect costs. A full cost analysis that identifies where the bulk of the costs lie (called a Pareto Analysis) and which categorizes costs into shop floor functions, can stimulate new ideas. Typical categories would be mechanical parts, electrical parts, assembly tasks, packaging, finishing costs. Set out in this way the

24 Value Analysis/Value Engineering

cost analysis can prompt several changes in direction. The following comments are typical of VA teams in action in the manufacturing and service sectors.

Table 1.7 Prompting ideas in the service sector

Functions	What functions does our service provide?	1. Can any function be eliminated? 2. Can any other way be found to achieve the same function? 3. Can another part of the service incorporate this function?
Service availability	What is the service availability?	1. Can any other service be provided? 2. Is the service over specified? 3. Can we provide the same service on a part-time basis?

Response time	What is the response reaction?	1. How is the service requested?
		2. Is the response rate inappropriate?
		   (a) By observation
		   (b) By comparison with associated functions and services
		   (c) By physical test
		   (d) By comparison with competitors' services
		3. Optimization problem – can any function be maintained with less cost or more functions at same cost?

Conversion of enquiries into sales	What percentage of enquiries are converted into provision?	1. Can pointless enquiries be identified earlier? 2. Can inappropriate enquiries be avoided by better advertising? 3. Can lost sales be avoided by improved marketing?
What are the key aspects of our service?	What are the primary, secondary and aesthetic parts of our service	1. Can any parts be eliminated? 2. Can any aspects be reduced? 3. Can we enhance our primary service? 4. Is it worth enhancing secondary and aesthetic functions?
Comparison with others	Which aspects of our service stand out?	1. Too good? 2. Relatively poor but tolerated? 3. Can we influence any associated problematic aspects? 4. What benefits are gained from offering our service – to recognized standards?
Service history	How do we deal with complaints	1. Short term response? 2. Long term response? 3. Comparison with competition?
Future services	What aspects of this service may change	1. Are we aware of improvements? 2. Could the service become obsolete? 3. Would any new functions in the service expand market share?

'We spend 40 per cent of the total on manual assembly. Why don't we completely re-design with automatic assembly as the paramount design parameter?'

Table 1.8 Service comments on 13 amp electrical plugs

Mechanical abuse of plugs
'The mechanical integrity of plugs cannot be ignored. In many workshop situations they are thrown on to concrete floors regularly and the whiplash effect of the lead can make impact speeds surprisingly high'.
'The self tapping retaining screw often strips the thread in the plug cap if over tightened. This also reduces electrical integrity'.
'If the main base is not sufficiently rigid it may be impossible to align pins with the outlet socket. People have been known to use steel cutlery to displace pins as they are pushed into sockets with disastrous results!'.
'The cable clamp is very often discarded by people who do not really appreciate its function'.
'Over zealous tightening of the cable clamp often causes the thread to strip and the function then becomes ineffective'.
'With a loose cover-retaining screw or a stripped thread, the plug cover can become detached whilst the plug is in a socket. Live metal parts are then exposed. Plugs are sometimes still used in this condition. If the live pin was attached to the cover this problem would be overcome'.

Electrical abuse
'The use of grub screws to attach wires to pins is adequate for low currents, but can cause high current densities on 2 or 3 kW equipment'.
'If any arcing occurs at grub screws, metal vapour can coat the inside of the plug and give current leaks to earth that trip earth leak detectors even though the equipment in use is fine'.
'We've seen all sorts of things used to replace fuses; nails, aluminium foil, hair grips. This is often because an electrical fault elsewhere has caused the supply of standard fuses to be used up. Unfortunately this now produces an electrical circuit that is capable of allowing massive currents to flow along unexpected paths'.
'The introduction of plastic shrouds to plug pins is to be applauded because kids often use steel cutlery to prise plugs out of sockets through curiosity'.
'Fuse holders need to be capable of lasting hundreds of fuse changes because it's much easier to keep changing fuses than to find the fault in the system'.

Miscellaneous
'Given the choice, many people request plugs with easy withdrawal loops on the caps even though they do not appear disabled or infirm'.

Table 1.9

Idea Evaluation Action Chart		Project: Electrical Plug	Date ....
Ideas	Advantages	Disadvantages	Action
1. Redesign cable clamp using viscoelastic material in a self-clamping arrangement	No screws needed. Fits any diameter cable.	If lead changed to thinner section clamp ineffective until shape is recovered. Already in use by competitor. May be patents applicable.	Find rapid response material and test design. Do patent search
2. Can live pin be fused or concealed so that cover loss means no live parts exposed?	Likely to be expensive. See if this can be introduced into new standard.	Long project involving attitude changes before payback.	Begin campaign for even safer plugs?
3. Can pins be plate steel? 4.	Cheaper.	Difficult manufacture due to thick plate needed for long term use.	Talk to coating experts.

'60 per cent of the product cost lies in 'bought-in' electrical components. Why don't we manufacture them ourselves or buy out a supplier?'

'70 per cent of the cost lies in the few specialist mechanical fasteners we make in small batch sizes. Why don't we redesign to eliminate them or to use cheaper standard devices?'

'30 per cent of our costs can be attributed to purchasing, storing and discarding food to maintain our extensive menu options; many of which are never ordered.'

The sort of analysis that prompts such comments is gathered on Cost analysis forms shown typically in Table 1.10. Once again this is based on the 13 amp electrical plug assembly shown in Fig. 1.4.

Such analyses are more difficult in the service sector. However, with some effort service provision can be broken down along these lines to reveal similar rationalizations.

Table 1.10

COST ANALYSIS FORM

Project *Electrical plug*   Date *13/3/95*

CATEGORY ANALYSIS

Item	No. off	Costs Material	Costs Labour	Total	%
Main base	1	10	–	10*	
Cover	1	4.0	–	4	
Cable clamp	1	1.0	–	↓	
screws	2	2.5	–	↓	
assembly	–	–	0.5	4	
Pins	3	7	–	↓	
grub screws	2	2	–	↓	
fuse plate	1	2	–	↓	
assembly	–	–	2	13*	
fuseholder	1	2	–	–	
grub screw	1	1.25	–	–	
terminal post	1	1	–	–	
assembly	–	–	0.75	–	
fuse (bought in)	1	2	–	7*	
Final assembly	1	1.25	0.75	↓	
+ screw				2	
GRAND TOTAL				40p	

		P %
Mechanical parts	15	37.5
Electrical parts	14.0	35
Assembly	4	10
Packing	0	0
Fasteners	7	17.5

Pareto Analysis

| 75 | % |

Cost in

| 3 |

items marked*

Project implementation

Once all the data and information has been collected and displayed in this way then the VA team begins the creative phase. It is inappropriate to give examples of the re-designs that occur in this phase because they are all specific to the project involved. The collection and analysis of ideas and data within a multi-discipline team is usually a sufficient catalyst to achieve product and service improvement. However the enthusiasm of the VA team does not always spread to senior management! Hence project conclusions should be presented in an organized and logical way that sets financial and technical findings alongside each other as follows.

Presenting VA conclusions

It helps if VA team findings are summarized on a Report Summary Form as shown in Table 1.11 for an improved electrical connection

28 Value Analysis/Value Engineering

system at the end of a tubular heater. The precise details are unimportant here but notice that all the relevant financial and technical information is displayed. Often several projects may be presented alongside each other for selection. If this is done on such forms it is easy to compare technical and financial data at a glance.

Table 1.11

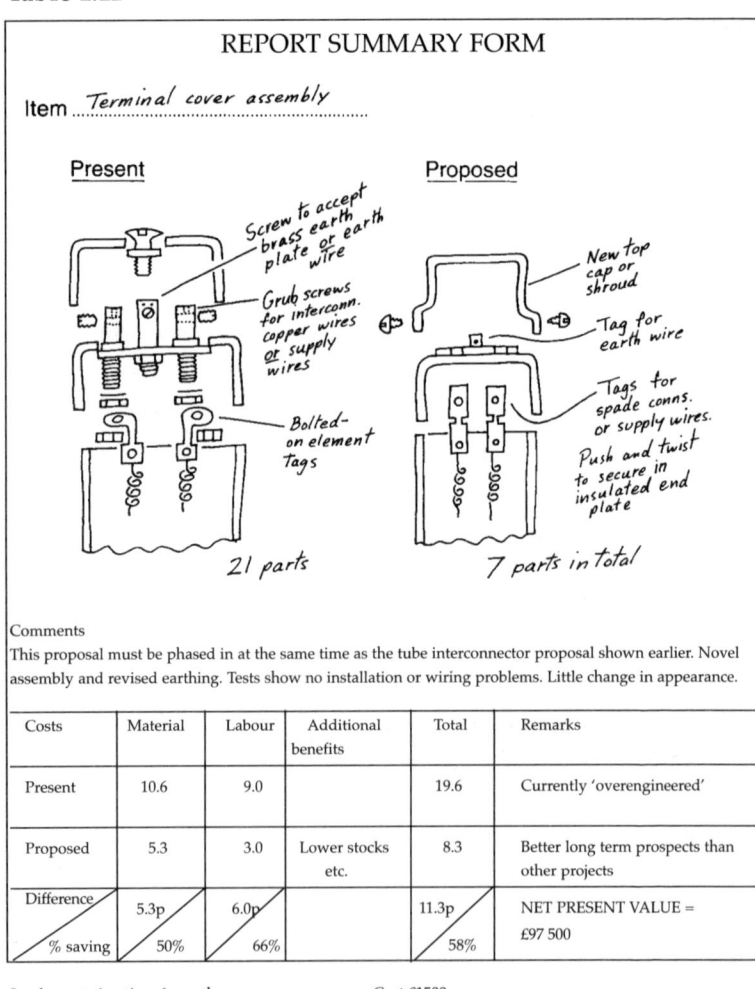

REPORT SUMMARY FORM

Item: Terminal cover assembly

Present (21 parts):
- Screw to accept brass earth plate or earth wire
- Grub screws for interconn. copper wires or supply wires
- Bolted-on element tags

Proposed (7 parts in total):
- New top cap or shroud
- Tag for earth wire
- Tags for spade conns. or supply wires
- Push and twist to secure in insulated end plate

Comments
This proposal must be phased in at the same time as the tube interconnector proposal shown earlier. Novel assembly and revised earthing. Tests show no installation or wiring problems. Little change in appearance.

Costs	Material	Labour	Additional benefits	Total	Remarks
Present	10.6	9.0		19.6	Currently 'overengineered'
Proposed	5.3	3.0	Lower stocks etc.	8.3	Better long term prospects than other projects
Difference / % saving	5.3p / 50%	6.0p / 66%		11.3p / 58%	NET PRESENT VALUE = £97 500

Implementation time 6 months
Quantity per year 140 000
1st year savings £~14 500

Cost £1500
Expected life 5yrs
Annual savings £~16 000

Value Analysis/Value Engineering   29

VA teams can get swept along by technically exciting projects with weaker financial prospects. Hence the purpose of such forms is to get them to face up to financial realities. This may soften the blow when technically exciting projects are turned down on financial grounds. Conversely technically exciting projects with less obvious financial benefit may be selected if all the facts are presented clearly. Occasionally such projects are chosen since the technical developments may give long term benefits that were not immediately obvious to the VA team but are part of wider corporate aims. A common conclusion is that several projects are technically equivalent and have to be compared on financial grounds. The next chapter addresses this in detail.

Objectives

After studying this chapter you should be able to do the following.

1. Give the approach used in Value Engineering and list the advantages and disadvantages in product and service improvement projects.
2. Show how projects are selected.
3. Demonstrate how a product or service can be considered as an assembly of functions and show how the functions can be ranked and costed.
4. Give examples of how ideas can be prompted.
5. Define or explain the following terms: value, function tree, primary function, secondary function, aesthetic function, Pareto analysis, Idea Evaluation Action Chart.
6. Describe why it is important to present Value Analysis findings in a standard format.

2 Financial appraisal methods for technically feasible projects

Aims

- To give appropriate methods by which the cost benefits of a project can be measured
- To introduce the concept of discounted cash flow and net present value in projects
- To give examples of how project plans can be costed for comparative purposes

Introduction

A common problem is that Value Analysis activities often produce several project plans that are considered to be technically and commercially equivalent. Hence any choice between which ones are embarked upon and which are shelved should be made on more detailed financial grounds. Almost all projects require some capital outlay and eventually produce some revenue. The rate of outgoings compared with cash returns is called 'Cash Flow'. Its optimization, within a 'free market' system where cash is always linked with an interest rate is the fundamental principle behind such evaluations.

Several methods are in common use to choose between projects

(a) Payback period
(b) Average gross annual rate of return
(c) Average net annual rate of return
(d) Discounted cash flow methods.

Methods (a)–(c) ignore the effects of time and interest rates on the value of money. They assume that a cash outflow early in the project has the same impact as a similar cash outflow at a later stage. Conversely they assume that income early in the project has the same value as income accrued much later. These methods therefore ignore the fact that cash is always linked with an interest rate; you are charged interest if you borrow the money needed or you could have gained interest on any money of your own by investing it elsewhere instead of tying it up in the project. Only method (d) takes interest rates and time into account. Hence it is a preferable method. However the simplicity of the first three methods means that they are sometimes used to eliminate projects that are non-starters in the initial phase. All methods ignore inflation, where the purchasing power of cash tends to depreciate with time. This fact and the problem that interest rates change with time in unpredictable ways means that projects often still require additional subjective assessment even when analysed by method (d).

The payback period

This requires you to calculate the initial capital investment into a project and then work out how long you need to recover the investment from the net cash flow income. Projects with shorter payback periods are favoured.

For example if you invest in a new press and tooling, total charges may come to £200 000 before production of injection mouldings. If you expect to accrue £40 000 per year clear profit from the sale of the mouldings then the payback period is

$$\frac{£200\ 000}{£40\ 000} = 5 \text{ years}$$

However sales can be influenced by various factors and Table 2.1 gives figures for two situations. In situation A, a massive initial advertising campaign is expected to give a rapid response to produce the net income shown. In situation B a planned refurbishment of the tooling is expected to give better net income towards the end of the project as shown. These costs are incorporated into the figures by calculating the net cash flow on an annual basis as shown.

32 Financial appraisal methods

Table 2.1

	Initial investment = £200 000 Levels of income	
Year	Situation A	Situation B
1	£80 000	£40 000
2	£60 000	£60 000
3	£40 000	£80 000
4	£30 000	£40 000
5	£30 000	£40 000
6	£30 000	£40 000

End of project, plant worn out, product obsolete.

Then the payback periods are:

Situation A = some time in the 4th year
Situation B = some time in the 4th year

hence choosing between the two situations would be tricky. Project A could be favoured because of the rapid return rates in the first two years. However project B appears to have a higher total revenue if the sales forecasts are accurate over the six year period. Clearly the limitations of the technique are showing up.

Average gross annual rate of return

This method concerns itself with the average gross income the project generates as a fraction of the initial investment. The figures are calculated for the expected life of the project.

For situation A (Table 2.1) assuming the same initial investment of £200 000, the total gross income is:

£270 000.

Hence over 6 years this is:

£45 000 per annum.

As a percentage of the initial investment this is

$$\frac{£45\ 000}{£200\ 000} \times 100 = 22.5\% \text{ gross return per annum.}$$

For project B the average income is

$$\frac{£300\,000}{6} \text{ per year}$$

= 50 000 per annum

= 25% gross return per annum.

Hence project B may be preferred under this type of analysis.

This type of assessment is generally applicable and used as a 'first filter' during project selection.

Average net annual rate of return

This method examines the net cash flow from the whole project once the initial investment has been recovered and averages it out over the life of the project. This average annual income is then compared with the average amount of investment in the project.

Once again for situation A in Table 2.1 the total income is £270 000 against the initial investment of £200 000. For calculation purposes it is assumed that £200 000 is set aside to recover the capital investment. Hence over the project life the total net income is £70 000.

The average **net** income per year for the six years is:

$$\frac{£70\,000}{6} = £11\,660 \text{ per annum}$$

For projects using mechanical equipment the investment is deemed to depreciate to zero over the project life and so the average level of investments is:

$$\frac{£200\,000 - £0}{2} = £100\,000$$

Hence the net annual rate of return is determined by comparing the average net income with the average investment level:

$$\frac{11\,660}{100\,000} \times 100 = 11.66\%$$

In situation B the net return is £100 000 over 6 years. Therefore the average net rate is £16 660 per annum.

34 Financial appraisal methods

Therefore the percentage net annual rate of return

$$= \frac{16\ 666}{100\ 000} = 16.66\%$$

Here situation B looks a much better project.

This type of comparison might be used in the plant and equipment hire industry to compare rental income with capital investment and plant life since it includes recovering the initial investment, writing off the equipment and being influenced by the total return on the investment.

For projects involving buildings and other assets that have an appreciable value at the end of the project it is often worthwhile incorporating it into the calculations. However, it should be borne in mind that industrial assets only achieve their potential value when they are associated with a thriving business. Hence the values of the investment should not be kept artificially high at the end of the project plan.

Discounted cash flow

This method takes into account a constant interest rate associated with the cash flows involved in the project. Such cash flows occur and can be assessed by compounding or discounting their values so that comparisons can be made at a fixed point in time.

Compounding

For example, if £1 is invested at 10 per cent interest and the interest left to build up with the initial investment, then the value of the total investment is seen to increase at 10 per cent each year as shown below:

Year 0 £1.00
Year 1 £1.10
Year 2 £1.21
Year 3 £1.33
Year 4 £1.46
and so on

Discounting

This is the same concept in reverse. Clearly if you had to meet a bill of £1.33 in three years time you could invest £1 at the present time at 10 per cent compound interest to meet the demand. Conversely the present value of £1.33 to be received in three years time is only £1. Hence the current value of £1 to be received in three years time is $\frac{£1}{£1.33}$ = £0.7513

Following this logic a discount table can be established as shown in Table 2.2 that gives the present value of £1 received or demanded n years into the future at various fixed interest rates. For example at 10 per cent interest rate, £1 to be received in six years is worth 56 pence **now** and a bill of £1 expected in three years time could be met by investing 75 pence **now**.

Net present value

In this method all cash flows are compared by discounting them to the present time, usually the project initiation point which we will call year zero. Outgoings are given a negative sign and net income is given a positive sign. Considering the projects in Table 2.1 the calculations for an interest rate of 10% are as shown in Table 2.3.

On this basis situation B would be favoured as the additional income in later years more than makes up for the poorer first year income. The net present value of all the cash flows is well over double that of situation A. You will recall that Situation B included a refurbishment in Year 3. If it cost £15 000 more than expected, you could accommodate this into the model by looking up the discount factor at 10 per cent interest at 3 years. As you will see this is 0.7513. Hence £15,000 could be generated by investing £(15 000 × 0.7513) now. This is –£11 269 NPV.

Hence the new value of £(20 788 – 11 269) = £9519 is still higher than situation A and so it still looks the favoured option.

Hence the discounting of the cash flows in projects enables them to be compared by assessing their net present values even though they follow different cash flow patterns. However, the successful outcome of a project still depends on achieving sales forecasts with known amounts of investment into the project. It also depends upon the interest rates remaining constant.

Table 2.2 Discount values

Interest rate		1%	2%	3%	4%	5%	6%	7%	8%	9%	10%	11%	12%	13%	14%	15%	16%
	1	0.9901	0.9804	0.9709	0.9615	0.9524	0.9434	0.9346	0.9259	0.9174	0.9091	0.9009	0.8929	0.8850	0.8772	0.8696	0.8621
	2	0.9803	0.9612	0.9426	0.9246	0.9070	0.8900	0.8734	0.8573	0.8417	0.8264	0.8116	0.7972	0.7831	0.7695	0.7561	0.7432
	3	0.9706	0.9423	0.9151	0.8890	0.8638	0.8396	0.8163	0.7938	0.7722	0.7513	0.7312	0.7118	0.6931	0.6750	0.6575	0.6407
Time	4	0.9610	0.9238	0.8885	0.8548	0.8227	0.7921	0.7629	0.7350	0.7084	0.6830	0.6587	0.6355	0.6133	0.5921	0.5718	0.5523
	5	0.9515	0.9057	0.8626	0.8219	0.7835	0.7473	0.7130	0.6806	0.6499	0.6209	0.5935	0.5674	0.5428	0.5194	0.4972	0.4761
in	6	0.9420	0.8880	0.8375	0.7903	0.7462	0.7050	0.6663	0.6302	0.5963	0.5645	0.5346	0.5066	0.4803	0.4556	0.4323	0.4104
	7	0.9327	0.8706	0.8131	0.7599	0.7107	0.6651	0.6227	0.5835	0.5470	0.5132	0.4817	0.4523	0.4251	0.3996	0.3759	0.3538
years	8	0.9235	0.8535	0.7894	0.7307	0.6768	0.6274	0.5820	0.5403	0.5019	0.4665	0.4339	0.4039	0.3762	0.3506	0.3269	0.3050
	9	0.9143	0.8368	0.7664	0.7026	0.6446	0.5919	0.5439	0.5002	0.4604	0.4241	0.3909	0.3606	0.3329	0.3075	0.2843	0.2630
	10	0.9053	0.8203	0.7441	0.6756	0.6139	0.5584	0.5083	0.4632	0.4224	0.3855	0.3522	0.3220	0.2946	0.2697	0.2472	0.2267
	11	0.8963	0.8043	0.7224	0.6496	0.5847	0.5268	0.4751	0.4289	0.3875	0.3505	0.3173	0.2875	0.2607	0.2366	0.2149	0.1954
	12	0.8874	0.7885	0.7014	0.6246	0.5568	0.4970	0.4440	0.3971	0.3555	0.3186	0.2858	0.2567	0.2307	0.2076	0.1869	0.1685
	13	0.8787	0.7730	0.6810	0.6006	0.5303	0.4688	0.4150	0.3677	0.3262	0.2897	0.2575	0.2292	0.2042	0.1821	0.1625	0.1452
	14	0.8700	0.7579	0.6611	0.5775	0.5051	0.4423	0.3878	0.3405	0.2992	0.2633	0.2320	0.2046	0.1807	0.1597	0.1413	0.1252
	15	0.8613	0.7430	0.6419	0.5553	0.4810	0.4173	0.3624	0.3152	0.2745	0.2394	0.2090	0.1827	0.1559	0.1401	0.1229	0.1079

Interest rate		17%	18%	19%	20%	21%	22%	23%	24%	25%	26%	27%	28%	29%	30%	31%	32%
	1	0.8547	0.8475	0.8403	0.8333	0.8264	0.8197	0.8130	0.8065	0.8000	0.7937	0.7874	0.7813	0.7752	0.7692	0.7634	0.7576
	2	0.7305	0.7182	0.7062	0.6944	0.6830	0.6719	0.6610	0.6504	0.6400	0.6299	0.6200	0.6104	0.6009	0.5917	0.5827	0.5739
	3	0.6244	0.6086	0.5934	0.5787	0.5645	0.5507	0.5374	0.5245	0.5120	0.4999	0.4882	0.4768	0.4658	0.4552	0.4480	0.4348
Time	4	0.5337	0.5158	0.4987	0.4823	0.4665	0.4514	0.4369	0.4230	0.4096	0.3868	0.3844	0.3725	0.3611	0.3501	0.3396	0.3294
	5	0.4561	0.4371	0.4190	0.4019	0.3855	0.3700	0.3552	0.3411	0.3277	0.3149	0.3027	0.2910	0.2799	0.2693	0.2592	0.2495
in	6	0.3898	0.3704	0.3521	0.3349	0.3186	0.3033	0.2888	0.2751	0.2621	0.2499	0.2383	0.2274	0.2170	0.2072	0.1979	0.1890
	7	0.3332	0.3139	0.2959	0.2791	0.2633	0.2486	0.2348	0.2218	0.2097	0.1983	0.1877	0.1776	0.1682	0.1954	0.1510	0.1432
years	8	0.2848	0.2660	0.2487	0.2326	0.2176	0.2038	0.1909	0.1789	0.1678	0.1574	0.1478	0.1388	0.1304	0.1226	0.1153	0.1085
	9	0.2434	0.2255	0.2090	0.1938	0.1799	0.1670	0.1552	0.1443	0.1342	0.1249	0.1164	0.1084	0.1011	0.0943	0.0830	0.0822
	10	0.2080	0.1911	0.1756	0.1615	0.1486	0.1369	0.1262	0.1164	0.1074	0.0992	0.0916	0.0847	0.0784	0.0725	0.0672	0.0623
	11	0.1778	0.1619	0.1476	0.1346	0.1228	0.1122	0.1026	0.0938	0.0859	0.0787	0.0721	0.0662	0.0607	0.0558	0.0513	0.0472
	12	0.1520	0.1372	0.1240	0.1122	0.1015	0.0920	0.0834	0.0757	0.0687	0.0625	0.0568	0.0517	0.0471	0.0429	0.0392	0.0357
	13	0.1299	0.1163	0.1042	0.0935	0.0839	0.0754	0.0678	0.0610	0.0550	0.0496	0.0447	0.0404	0.0365	0.0330	0.0299	0.0271
	14	0.1110	0.0985	0.0876	0.0779	0.0693	0.0618	0.0551	0.0492	0.0440	0.0393	0.0352	0.0316	0.0283	0.0253	0.0228	0.0205
	15	0.0949	0.0835	0.0736	0.0649	0.0573	0.0507	0.0448	0.0397	0.0352	0.0312	0.0277	0.0247	0.0219	0.0195	0.0174	0.0155

Table 2.3 NPVs for the projects in Table 2.1

	Year	Net cash flow	Discount factor	Net present value
Situation A	0	−200 000	1.0	−200 000
	1	+80 000	0.9091	72 728
	2	+60 000	0.8264	49 584
	3	+40 000	0.7513	30 052
	4	+30 000	0.6830	20 490
	5	+30 000	0.6209	18 627
	6	+30 000	0.5645	16 935
				+8416
			Situation A = net present value	
Situation B	0	−200 0000	1.0	−200 000
	1	+40 000	0.9091	36 364
	2	+60 000	0.8264	49 584
	3	+80 000	0.7513	60 104
	4	+40 000	0.6830	27 320
	5	+40 000	0.6209	24 836
	6	+40 000	0.5645	22 580
				+20 788
			Situation B = Net present value	

Applications of NPV techniques

So far we have considered examples where investments generate net income over a number of years. The examples given above considered capital investments in year zero and in Situation B the second major expense of a refurbishment in year three was also included.

The technique can also be used to assess advertising campaigns. Suppliers of products and services are faced with many possibilities to advertise their businesses. Some may be inappropriate, but usually, there are a few feasible plans. Different methods have different initial costs, repeat costs and response rates. The NPV method can be used if accurate predictions for the additional net income generated can be made. Use the discount table to assess the NPV of the campaigns detailed in Table 2.4 to promote the sale of a new tennis racket assuming an interest rate of 5 per cent prevails throughout the project.

Financial appraisal methods

Table 2.4 Tennis racket promotion and income details

Campaign A	Campaign B	Campaign C
Series of TV adverts	Player endorsement on 6 year contract	Newspaper and magazine campaign
Year	Year	Year
0 – £1.5 M	0 – £2 M	0 – £1 M
2 – £0.5 M	1 – £2 M	1 – £0.5 M
	2 – £1 M	2 – £0.4 M
	3 – £0.5 M	3 – £0.3 M
	4 – £2 M	4 – £0.5 M
	5 – £1 M	5 – £0.4 M
		6 – £0.3 M
		7 – £0.1 M
	Additional net income predicted	
Year	Year	Year
1 + £3.0 M	1 + £3 M	1 + £2 M
2 + £0.7 M	2 + £2 M	2 + £2 M
3 + £1.5 M	3 + £1 M	3 + £0.8 M
4 + £0.5 M	4 + £2 M	4 + £0.9 M
5 + £0.2 M	5 + £0.5 M	5 + £0.6 M
6 + £0.1 M	6 + £0.1 M	6 + £0.3 M
7 negligible	7 negligible	7 + £0.1 M

As can be seen from the answer given at the end of the chapter the TV campaign gives the best NPV and the player sponsorship/endorsement actually loses money. Manufacturers often spread their budgets throughout the possible avenues but the same calculations could be done for the different ways in which the budget could be divided.

However before revenue can be obtained from any chosen project there are many other hurdles that have to be jumped during the implementation phase. The next chapter deals with the timing of events during the life of a project.

Answer to tennis racket advertising campaign

The NPVs of the three separate campaigns have been calculated using the discount table and are given in Table 2.5.

Table 2.5 NPVs for the tennis racket campaigns

Year	Discount factor	Campaign A		Campaign B		Campaign C	
		Cash flow	NPV	Cash flow	NPV	Cash flow	NPV
0	1.0000	−1.5	−1.500	−2.0	−2.000	−1.0	−1.00
1	0.9524	+3.0	+2.857	+1.0	+0.952	+1.5	+1.429
2	0.9070	+0.2	+0.181	+1.0	+0.907	+1.6	+1.451
3	0.8638	+1.5	+1.296	+0.5	+0.431	+0.5	+0.432
4	0.8227	+0.5	+0.411	0	0.000	+0.4	+0.329
5	0.7835	+0.2	+0.157	−0.5	−0.391	+0.2	+0.157
6	0.7462	+0.1	+0.075	+0.1	+0.075	0	0.000
7	0.8107	0	0.000	0	0.000	0	0.000
	NPV totals		£3.477M		−£0.026M		£2.798M

Objectives

After studying this chapter you should be able to do the following:

1. Give four ways of appraising projects in terms of the cash flows involved.
2. Explain how time and interest rates can be used to discount cash flows and give any project a net present value.
3. Calculate the net present value of a project from cash flow data and a discount table.

3 Using Network Analysis to implement projects

Aims

- To demonstrate how projects are composed of interacting achievements within the project framework
- To show that there will be a critical path, through the project achievements, in which delays above a certain level will definitely delay the project
- To show how a Network Analysis can influence cash flow and logistics decisions in projects.

Introduction

After value analysis and financial appraisal comes the task of implementing the selected project. This will usually mean organizing many smaller tasks so that they mesh together as quickly and effectively as possible. It is also useful to identify key stages when specific tasks **must** have been completed. For instance it is pointless inserting late-delivery penalty clauses into the specification of an injection moulding tool if it cannot be used as soon as it is delivered. This might be because the press on which it is to be used, has not yet been delivered. Worse still it may be because the concrete base has not been completed. Such planning should also be linked with the financial planning because if you can order things as late as possible, you can pay for them as late as possible and improve the cash flow situation.

With large projects the number of individual tasks can get daunting. However we will consider a system that can, in principle, be extended to cover any size of project.

Using Network Analysis to implement projects

The idea is to turn the project into a series of events that are represented by arrows. The event is shown above the arrow and the minimum duration possible shown below. Unlike vector analysis the physical length of the arrows has no real meaning. Tooling manufacture that must take 51 days minimum is represented in Fig. 3.1.

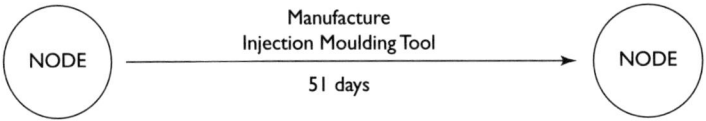

Figure 3.1 A network event

Events begin and end at nodes which will eventually show the information in Fig. 3.2 within the node circles.

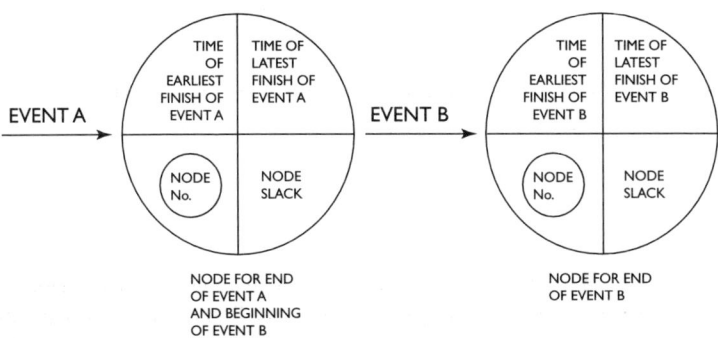

Figure 3.2 The essential information in nodes

The **node slack** is the difference between the latest time the event could be completed without creating a delay in the remainder of the project and the earliest possible time the event could finish if **all** previous associated events are achieved on schedule. As you will see later, nodes with the lowest slack identify critical events in the project plan.

So for boiling an egg there are two sequential events:

A   Boil water – 6 MINUTES
B   Cook egg – 4 MINUTES

42 Using Network Analysis to implement projects

If you are allowed 12 minutes to do the job the events would be networked as shown in Fig. 3.3.

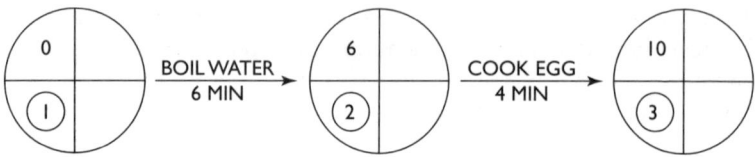

Figure 3.3 The foundations of a Network Diagram for egg boiling

So far we have inserted the earliest possible finish times at each node by working left to right and adding the event times. Notice that the sequence starts at time 0 in the far left node and the whole task could be done in 10 minutes.

If we now insert 12 minutes as the latest possible finish time (the time we've been allowed) and work backwards we see that the start could have been delayed by 2 minutes (Fig. 3.4).

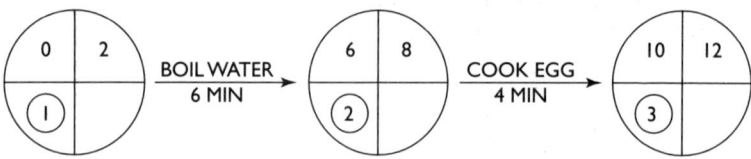

Figure 3.4 Latest finish times added

We can now work out the slack at each node by subtracting the value in the top left hand quadrant from the top right hand quadrant (Fig. 3.5).

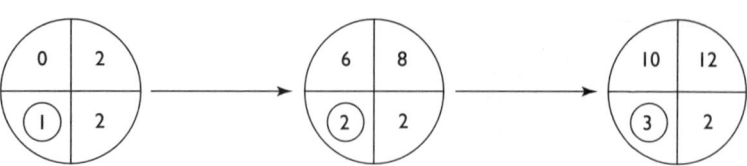

Figure 3.5 Node slack calculated

Using Network Analysis to implement projects

As you can see there is two minutes slack in the system which could be utilized at any of the nodes. However because eggs keep cooking even when they are removed from pans and it costs money to keep water boiling you would probably use up the slack by delaying the start by 2 minutes if you are certain that the water can be boiled in 6 minutes. You will see later how such practicalities can be used to fine-tune more complex networks.

Constructing networks

When several events have to occur before another can start the diagrams are modified as shown in fig. 3.6. This typical sequence of events represents the installation and commissioning of an injection moulding station.

Construct Press (6 months)
Make Moulding Tool (3 months)
Lay Foundations (1 month)
Commissioning Trials (1 month)
Pilot Production Run (0.5 month)
Target Completion Time (12 months)

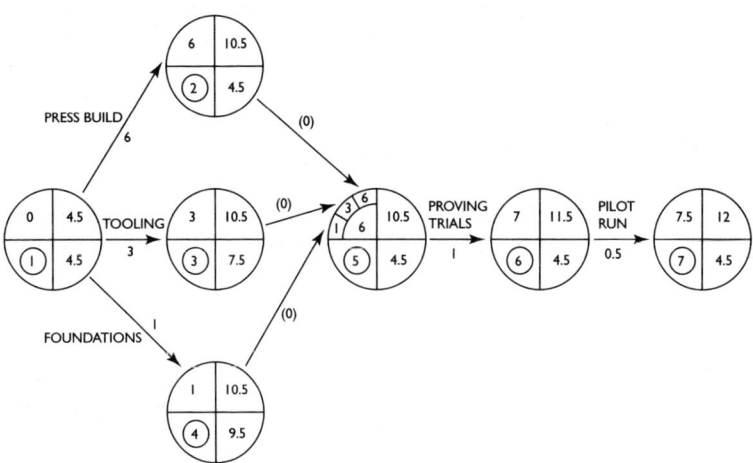

Figure 3.6 A network diagram for an injection moulding station

As you will see it has been necessary to introduce three **dummy events** of zero duration before node five to represent the fact that the proving trials cannot begin until the press, tooling and foundations have been completed. For ease of logic, node five has been modified to display the three possible earliest start dates for the next event. Obviously the largest value must be transferred into the relevant inner quadrant since it represents the time at which **all** essential prior events have just been finished.

Notice too that a critical path, where the lowest level of slack occurs is the path between nodes one, two, five, six and seven. For this reason the technique is also known as **critical path analysis**.

At present the slack in the system is 4.5 months. The slack in this path could be reduced to zero if the target completion date was reduced by 4.5 months. In such circumstances special attention would be made on all events along the critical path to keep the project on schedule.

It doesn't take much to complicate things. Consider this project. A family has to prepare, cook and eat a 2 course meal of meat and vegetables followed by pie and custard within two hours. The events take the times shown in Table 3.1.

Assuming that there is adequate labour and equipment to start as many events as possible simultaneously, the network diagram would be laid out as shown in Fig. 3.7.

Table 3.1

A	Defrost meat	20 mins
B	Prepare vegs	15 mins
C	Prepare pie	20 mins
D	Bake pie	45 mins
E	Make custard	10 mins
F	Lay table	5 mins
G	Cook vegs	15 mins
H	Cook meat	50 mins
I	Serve 1st course	3 mins
J	Eat 1st course	20 mins
K	Eat 2nd course	10 mins

Using Network Analysis to implement projects 45

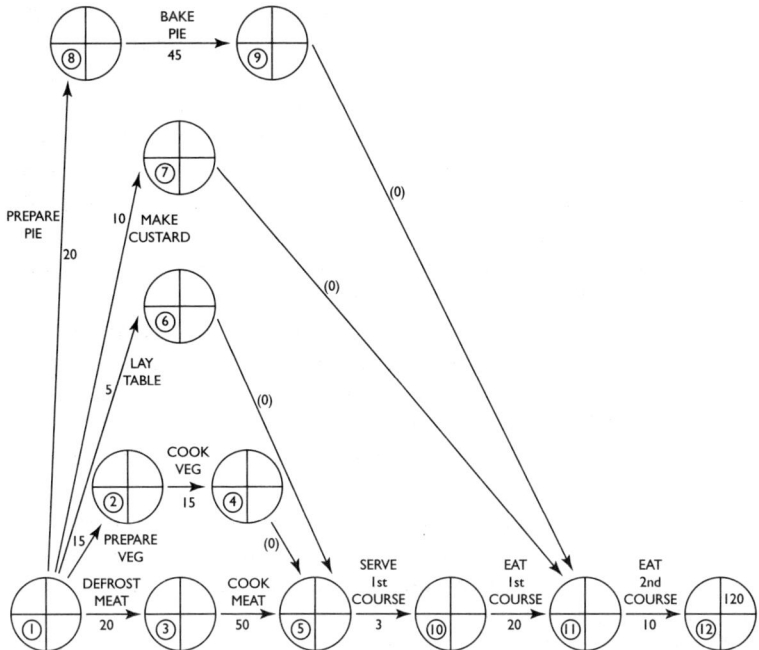

Figure 3.7 Outline network diagram for 2-course meal

Notice that four dummy events are included to formalize the sequence of events. The completed diagram is shown in Fig. 3.8

Clearly the critical path carries a slack of 17 minutes which could be removed by bringing the finish time forward to 103 minutes. Notice that the other paths still contain considerable slack.

Excessive slack in non-critical paths can cause trouble in projects. It might mean, for instance, that materials and equipment arrive on site long before they are needed with consequential environmental and security problems. It might also mean that documents and quotations are prepared too early and become out-of-date.

46 Using Network Analysis to implement projects

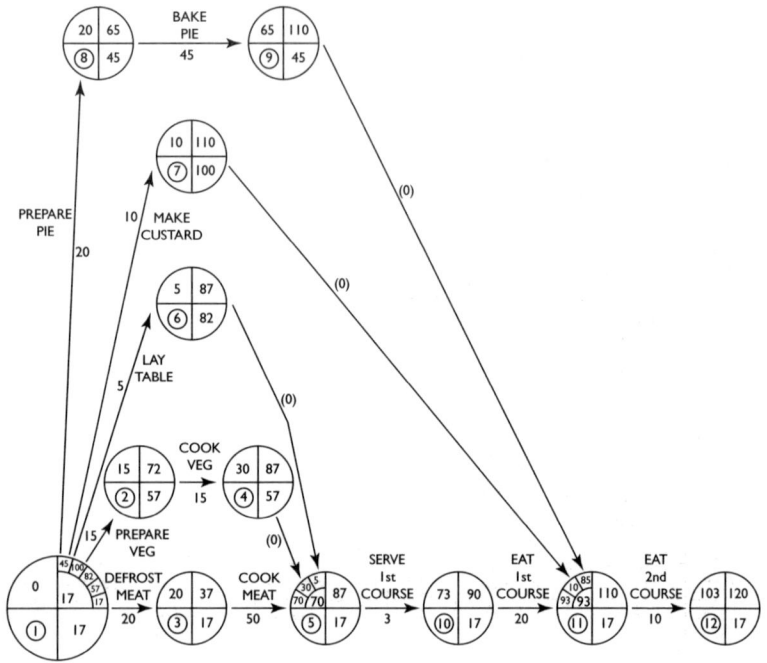

Figure 3.8 Network diagram for 2-course meal (completed)

Optimizing delays, working patterns and cash flows in projects

In our domestic example we might wish to ensure that the pie and custard arrive freshly prepared and baked and that the table is laid at the last possible moment. These events would also make better use of the labour available and ease traffic in the kitchen at the start of operations.

This is done simply by introducing **delay events** by working backwards along the paths and introducing delays to take up the slack at the start or at any other convenient point. This is done in Fig. 3.9 for all but one of the loops. Notice too that the critical path now contains zero slack.

Using Network Analysis to implement projects 47

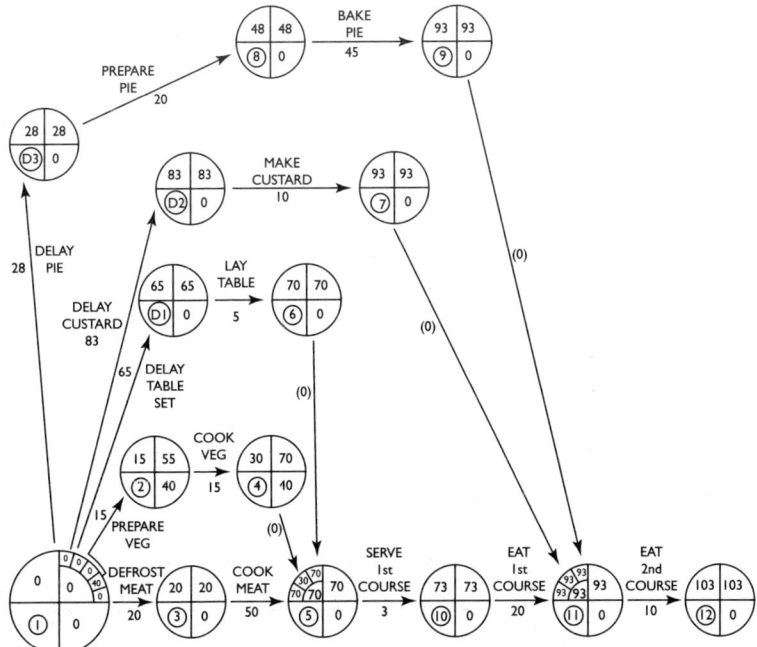

Figure 3.9 Complete diagram with delays and zero slack in critical path

As you can see there is still slack in the path involving the vegetables. Perhaps you might work out the best point to delay this path to ensure the vegetables are freshly cooked at node **5**. This has been done in Fig. 6.2.

Delay events should be positioned where they are most beneficial. If they are part of a loop that involves cash flows out of the project then their best use is to delay the outflow. This means ordering equipment at the latest possible time. This can be represented by an event called 'place order' of zero duration. If demarcation of labour skills is not a problem then delays can also be used to spread the tasks out within the project to minimize the need to allocate additional staff to tasks that are occurring simultaneously. Where specialist tools, equipment and personnel are in limited supply, tasks which require them should be staggered to optimize their use.

48 Using Network Analysis to implement projects

Conversely events that bring in revenue should be placed as early as possible together with an event that ensures all invoices are sent out as soon as possible.

Such techniques are clearly related to 'Just-in-Time' production and delivery schedules. If you know the production rates that your external or internal suppliers can work to and you know your requirement then you know the minimum duration that a production request should take. This can therefore be included as a 'supply task' of known minimum duration and placed as late as possible in a loop with an appropriate slack.

If you have to pay extra to get late delivery penalty clauses in contracts then you are recommended to use such resources in the most critical path.

Objectives

After reading this chapter you should be able to:

1. Draw a project network diagram for lists containing sequential events and their durations.
2. Identify the critical path within the network diagram.
3. Inspect the diagram and make recommendations on how to improve cash flow and logistics events such as ordering policy, security and workloads within the project.

4 Optimizing production rates and working practices

Aims

- To demonstrate that the optimum rate of working for equipment is not always the one that gives the fastest production rate.

- To demonstrate that the optimum working pattern for the workforce in a factory depends on many factors including the total output level, the rates of pay for direct workers and the rates of pay for indirect workers.

- To demonstrate that similar concepts to those above can be used in the service sector to determine the optimum 'level of service'.

Introduction

Once a new project has come to fruition and the new production process or service is fully commissioned it is then necessary to optimize the working practices for the duration of the task. The best long term working parameters are not necessarily those that give the highest initial production rates, or their equivalent in the service sector. There are many reasons for this and some are offered in the sections that follow.

Premature attrition of plant, equipment and tooling

In manufacturing a batch of size N_b components, factory time is spent as:

50 Optimizing production rates and working practices

(1) Non-productive time t_n per component (loading, warming up equipment, etc.)

(2) Productive time per component t_p.

(3) Average breakdown and repair time t_x (tool changing, resetting, etc.).

Thus the total time spent on the batch is

$$(N_b t_n + N_b t_p + N_t t_x)$$

Where N_t is the number of breakdowns or tool resets etc. for the batch.

If the operating costs of the factory are M/unit time then the 'factory time' cost of a batch N_b is

$$M(N_b t_n + N_b t_p + N_t t_x)$$

To this must be added the additional cost of the breakdowns. This includes new tools and moulds or repair costs, etc. If C_x is the average additional cost of each breakdown then this batch of N_b will attract a breakdown cost of $N_t C_x$.

Therefore the total costs of this batch of size N_b

$$= MN_b t_n + MN_b t_p + MN_t t_x + N_t C_x$$

Therefore the total production costs C_p per component is obtained by dividing by N_b (batch size). Hence

$$C_p = Mt_n + Mt_p + \frac{N_t M t_x}{N_b} + \frac{N_t C_x}{N_b} \qquad \text{(Equation 4a)}$$

The first term is the non-productive cost due to the necessary unloading, equipment warm-up and other indirect times that are associated with the particular process and plant and are unavoidable.

The second term is the essential production cost and can be minimized by reducing the production time. Ideally this could be minimized by driving the production process as fast as it could go.

The last two terms relate to the costs of refurbishing the production system. First there is the cost of the downtime at the appropriate rate (M) and second the intrinsic cost of new or refurbished equipment when it is required.

Let's now concentrate on the second term. There are many instances where driving equipment too hard gives an accelerating rate of attrition. The best example comes from the metal working industry where machining tool tips, wire and tube drawing dies, extrusion dies, shearing blades and many other items that are in direct contact with the metal, wear out in a disproportionate and accelerating fashion if the rate of working is increased. This is usually attributed to the higher temperatures generated at the interface and the higher forces needed to perform the task at high speed.

The relationship is as shown in Fig. 4.1 for a metal cutting tool.

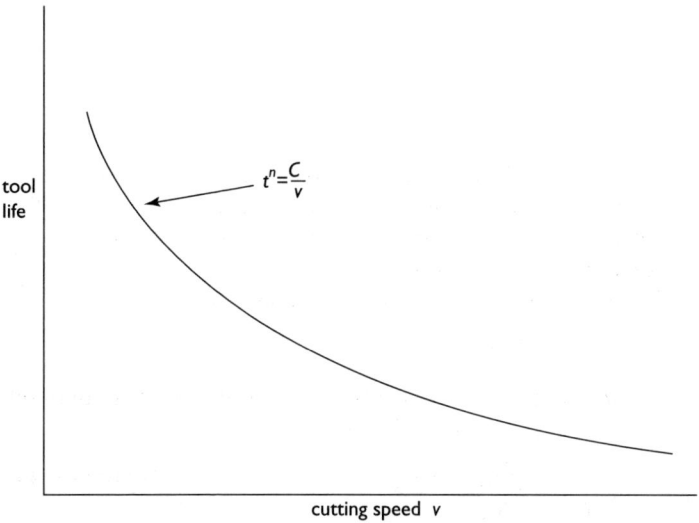

Figure 4.1 Effects of increasing cutting speed on tool life for otherwise constant machining conditions (feed and depth of cut)

Now the number of tools needed for the batch of components (N_t) is $N_b t_p / t$ where t is the average expected tool lifetime under the prevailing condtions. Hence

$$\frac{N_t}{N_b} = \frac{t_p}{t} \qquad \text{(Equation 4b)}$$

but $t^n = \frac{C}{v}$ (from Fig. 4.1) where C is a constant for the particular conditions.

52 Optimizing production rates and working practices

$$\therefore t = \left(\frac{C}{v}\right)^{\frac{1}{n}}$$

and substituting this into Equation 4b gives

$$\frac{N_t}{N_b} = \frac{t_p}{\left(\frac{C}{v}\right)^{\frac{1}{n}}} = \frac{t_p x v^{\frac{1}{n}}}{C^{\frac{1}{n}}}$$

but $C^{\frac{1}{n}}$ is just another constant because n is a constant, let's call the new constant C_2

$$\therefore \frac{N_t}{N_b} = \frac{t_p v^{\frac{1}{n}}}{C_2} \qquad \text{(Equation 4c)}$$

Now during machining the actual machining time is the productive time t_p and is given by the distance over which the tool contacts the workpiece divided by the speed of the tool relative to the workpiece. For a cylindrical turning operation this is approximately the circumference multiplied by the number of times the workpiece rotates during the operation. Let's generalize and call the distance K.

$$\therefore t_p = \frac{K}{v} \qquad \text{(Equation 4d)}$$

substituting equations 4c and 4d into 4a gives

$$C_p = Mt_n + MKv^{-1} + \frac{Kv^{\frac{1}{n}}}{vC_2}(Mt_x + C_x) \qquad \text{(Equation 4e)}$$

$$C_p = Mt_n + MKv^{-1} + \left(\frac{K}{C_2}\right)(Mt_x + C_x)v^{\frac{1}{n}-1}$$

when v is small the second term is big and dominates. When v is big the third term is big and dominates. Hence C_p falls through a minimum somewhere between these extremes as shown in Fig. 4.2.

Differentiating Equation 4e with respect to v gives

$$\frac{dC_p}{dv} = -1MKv^{-2} + \left(\frac{1}{n}-1\right)\left(\frac{K}{C_2}\right)(Mt_x + C_x)v^{\frac{1}{n}-2}$$

setting this equal to zero will establish the value of v that gives the minimum unit cost C_p

$$0 = -1MKv^{-2} + \left(\frac{1}{n}-1\right)\left(\frac{K}{C_2}\right)(Mt_x + C_x)v^{\frac{1}{n}-2}$$

$$\therefore MKv^{-2} = \left(\frac{1}{n}-1\right)\left(\frac{K}{C_2}\right)(Mt_x + C_x)v^{\frac{1}{n}-2}$$

$$\therefore MK = \left(\frac{1}{n}-1\right)\left(\frac{K}{C_2}\right)(Mt_x + C_x)v^{\frac{1}{n}}$$

$$\therefore v = \left(\frac{MC_2}{\left(\frac{1}{n}-1\right)(Mt_x + C_x)}\right)^n$$

$$\therefore v = \left(\left(\frac{n}{1-n}\right)\left(\frac{MC_2}{Mt_x + C_x}\right)\right)^n$$

Notice how the optimum speed for minimum total cost per component now depends upon the cost rate of running the factory (M), the time taken when a breakdown/repair occurs (t_x), the additional costs of the breakdown (C_x) and the very strong dependence the sensitivity of wear rate with cutting speed (n). Notice too how the actual length of the cut (K) cancels itself out in the analysis.

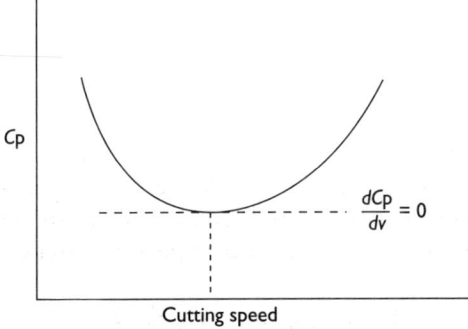

Figure 4.2 Production cost as a function of cutting speed

In most situations the average rate of expenditure to run the operation (M) is known. The breakdown times and the costs of refurbishment can be found out. The constants n and C_2 are usually the most difficult to establish. In this situation they can be found as follows:

$$t^n = \frac{C}{v} \text{ and } C_2 = C^{\frac{1}{n}} \quad \text{(given before)}$$

taking logs of the first equation gives

$$n \log t = \log \left(\frac{C}{v}\right)$$

$$\therefore n \log t = \log C - \log v$$

$$\therefore \log v = \log C - n \log t$$

If several tool life estimates are made at different speeds and a plot of $\log v$ against $\log t$ is performed it is comparable to the equation of a straight line

viz: $y = c - mx$ (see later in Chapter 5)

where c is the intercept on the y axis when x is zero and $-m$ is the gradient.

Hence the machining trials should plot out as shown in Fig. 4.3 if the proposed relationship is valid.

Hence n and C can be determined and from these C_2 can be established. Now all the information is known to calculate the optimum machining rate for the situation.

Using the approach in other situations

Clearly the approach can be easily transferred to other metalworking equipment. However before it can be used in other situations the equivalent relationship needs to be established. Shown in Fig. 4.4 is a **schematic** graph that may be applicable to the equipment-hire industry where servicing the equipment has a cost and reduces the time it is available for hire. It is only done because it extends the lifetime of the equipment.

Hence if hire companies can establish the precise form of this relationship they could optimize their hire revenue against the investment made, by servicing at the optimum interval. This may

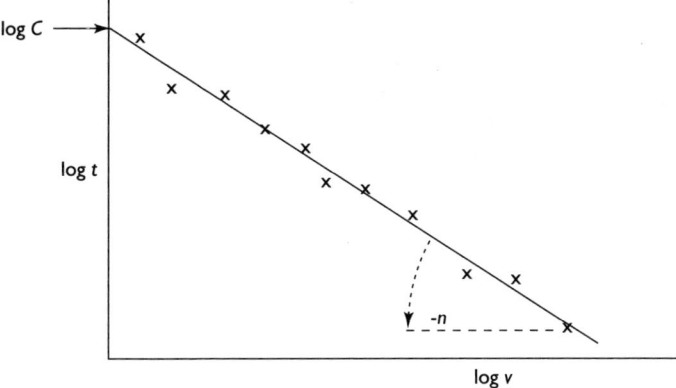

Figure 4.3 Machining trials, tool life t as a function of cutting speed v

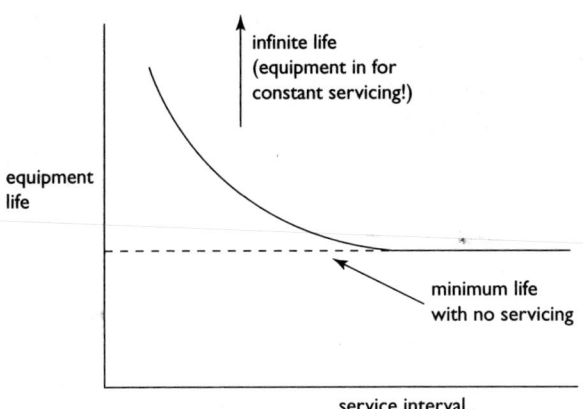

Figure 4.4 Equipment lifetime – service interval schematic for plant hire considerations

56 Optimizing production rates and working practices

differ from the manufacturers' recommendations. It would be shorter if large hire fees can be charged and longer if only modest hire fees could be commanded.

In the service sector a balance between numbers of sales staff and 'customer care' has to be achieved at points-of-sale. This balance occurs all the time at check-outs in supermarkets, pay kiosks in cafeterias, box-offices, bank clerk stations and many more situations.

Here, you might measure 'customer pressure' in terms of the length of wait they have to expect, against 'customer satisfaction'. This might be plotted as shown in Fig. 4.5.

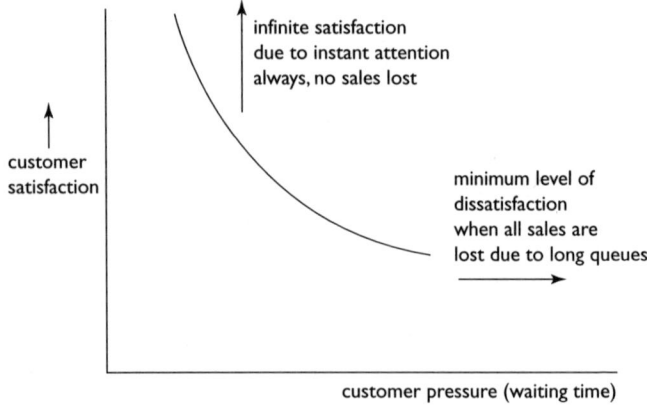

Figure 4.5 Point of sale relationship

Unless you are selling tickets to an oversubscribed blockbuster then you are likely to lose business when customers have to wait very long periods of time for service. Conversely if you provide so many sales personnel that queues never build up then you also may not be commercially optimized. The tolerable wait also depends on the value of the sale compared with the salary of the sales person. Hence sales staff sit around in luxury-car dealerships and supermarket check-out assistants rarely have clear gangways. Therefore if you can establish the loss of business associated with certain levels of customer satisfaction then the above method could be used to establish the optimum customer pressure for your service situation.

With all these methods the aim is to find an equation between the total cost and the key variable. In the machining case it was the cutting speed v. If this equation can be established it may be possible to differentiate it. Hence all dependent variables need to be expressed in terms of the key variable. This was possible in the cutting case. If the equation cannot be differentiated then the minimum can still be found by graphical means.

Non-recovery of tooling plant and equipment costs

Contrary to the previous section there are some situations where tooling wears out relatively slowly and may become obsolete long before the end of its economically viable life. In such cases the initial costs may not be fully recovered. This is exemplified by the launch of a new product containing complex injection mouldings. Toy production is typical. The rate at which mouldings can be removed from such tools depends on the rate at which the heat can be conducted through and away from the plastic part. At very high production rates it is the thermal conduction through the plastic that is the limiting factor and little can be done to speed this up. Hence, for products that are linked to fashionable and seasonal trends or associated with advertising initiatives, the 'polymer converters', who operate such tools, are under considerable pressure to duplicate tooling or to commission more expensive multiple cavity moulds in order to satisfy immediate demand. If they have adopted a policy to recover tooling costs over the engineering life of the mould, measured in terms of product numbers produced, it is essential that each tool runs to the end of its life. If demand suddenly ceases, the owners can be left with very expensive, part worn, obsolete tooling on which the initial costs have not been recovered.

There are several ways to counter this problem. A common way is to insist that the client purchases the tools at the outset of the project or to introduce penalty clauses if the full production run is not taken up. However penalty clauses are difficult to follow up with companies that are in financial difficulties. Hence agreements where initial tooling costs are shared is a common way of sharing commercial risk. It is such factors that make projects with short pay-back times attractive, even though there may be longer term projects with better Net Present Values. (See Chapter 2).

Such problems can also be encountered in the extreme when customers demand components at short notice or require very high but intermittent production rates to satisfy 'just-in-time' parts supply. A supplier can be pressurized into the duplication of tooling and plant if it has not been possible to build up buffer stocks because local accountancy and/or facilities do not allow for significant stock holding. If supply continues at such levels then all is fine. However if demand falls or modifications are introduced that make tooling obsolete then the supplier is left with half-used tools.

Under-utilization of resources in the service sector

Similar problems are encountered in the service sector when projects require large initial investments in order to initiate projects. There are often considerable third party risks that are beyond the control of the primary project team. The following examples highlight the types of problems that may occur:

1. Service provisions that are contracted out by larger project ventures. For example restaurants, transport systems, cleaning contracts, maintenance contracts, etc., set up by large 'theme park' ventures. Many such subsidiary projects require accurate long-term predictions of customer visits to the major venture to justify the initial investment. If visitor levels fail to reach planned targets there are serious repercussions to the subcontractors.

2. Franchise agreements that fail. Here a major supplier might require capital inputs from many smaller ventures in order to enter the market. Ideally the franchisee benefits from economies of scale generated by the franchiser for the purchase of the service material or product. These economies of scale also generate advantages on advertising campaigns and sales initiatives. There is also the possibility of market control in order to prevent over-saturation in any geographical area.

   However all these advantages can disappear if the umbrella organization fails or has overestimated the market potential. This can leave the franchisees with impossible commercial

targets relative to the investment made. Hence franchise and sub-contract agreements require very careful scrutiny.
3. Inadequate long-term market investigations can also create problems. The acquisition of the garage and petrol station in the town that is about to be by-passed is the classic example! There are many other examples of projects that fail due to incomplete enquiries and unfortunate developments. The fashion, music and entertainment industries are obvious minefields.

Once again these risks can be minimized by adopting a Value Analysis approach to the projects. The multi-discipline approach involved has the best chance of foreseeing such changes to legislation, technology and practices that may jeopardize the project

Optimization of fixed and variable costs

It is a fact of industrial life that the variable (direct) costs of commissioning and bringing a project into action must be done within a safe and controlled environment. This means that the direct costs of a project are not the only concerns. The fixed (indirect) costs associated with the provision of the working environment must also be taken into account. In Chapter 2 the cash flows considered for the projects had already taken this into account. However it is essential to consider in detail how direct and indirect costs are combined. Variable costs are usually easy to establish because they represent the minimum needed to achieve the task and include materials, wages of direct workers and other costs that are completely attributable to the production of a product or the provision of a service.

Once the variable costs have been established several methods are used to allocate the fixed costs.

(1) Increase the variable costs by a known percentage. This percentage can be established from historical records from the organization or from comparable records if no local records are available. This percentage varies enormously depending upon the precise nature of the product and service. It also depends on the size and nature of the organization. Small organizations tend to have lower overheads but suffer from not being able to

generate economies of scale on the variable costs. This method gives a 'blanket' approach and is insensitive to variations within the location.

(2) Break down the costs of all tasks into fixed and variable costs. The fixed costs are associated with the provision of the opportunity to perform each task. Hence they are not strongly related to the level of production. The variable costs are then the remaining costs that do depend on the level of activity. Hence material costs are wholly 'variable', but the salaries of the buyer and storekeeper who are responsible for acquiring and issuing the material are fixed costs. The final product or service cost is then achieved by adding the costs of all contributing tasks together. Clearly this method is much more cumbersome than the first method. However it can give an exact assessment of where overall expenditure occurs and can expose local variations in the proportions of fixed and variable costs.
(3) A compromise is to try and establish overhead percentages for sub-units within an organization. These recognize that certain activities attract different levels of fixed costs. This method usually makes use of past performance data collected on a sub-unit basis and uses the figures to apportion future fixed costs. For instance a new automated production line would incur a lower overhead charge than one where much maintenance, tool setting and inspection was required.

It is now assumed that fixed and variable costs can be determined for each unit of production or service task at a certain level of production. However the total of these two costs depends on the level of production. In successful times it is often tempting to increase production levels or service provision towards the absolute maximum that the factory can achieve. Hence, overtime, weekend working, second shift and round-the-clock production is a common sequence of events in an expanding business. However very careful consideration is needed to establish the optimum response to a rising market. This is particularly true when the demands fluctuate even though there is a general increase with time. The major problem is that the fixed costs can get out of proportion during minor changes in production level.

If the unit costs of production are calculated by combining fixed and variable costs they will fall with increased levels of production until the maximum output from one eight hour shift has been achieved. This is for two reasons. First, the relative contribution of the fixed costs falls and second, economies of scale should be obtainable on variable costs as absolute levels of production rise. This effect is shown in Fig. 4.6. The curve relates to a rise in production levels in a factory operating a five day per week, eight hour shift per day, working pattern. A 40 hour production period only occupies 24 per cent of the total of 108 hours in a full week. Hence the curve ends as shown.

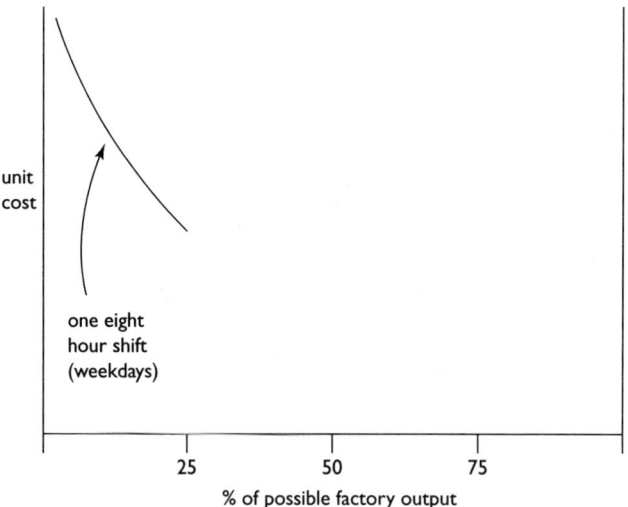

Figure 4.6 Unit costs of production as a function of production activity levels

If demand rises above 24 per cent then a decision on how to achieve this increased output has to be made.

The choices are

(1) weekday overtime working,
(2) weekend working,
(3) introduction of a second shift.

The unit costs for items produced by this additional work may be represented as shown in Fig. 4.7.

As can be seen the extra variable costs of production at overtime rates results in an increase in unit production cost. However these will decrease slightly with increased activity. There is also a minor increase in fixed costs if security men and other indirect workers are needed during the overtime hours.

Clearly the amount of overtime that can be tolerated by the workforce is limited as indicated on Fig. 4.7.

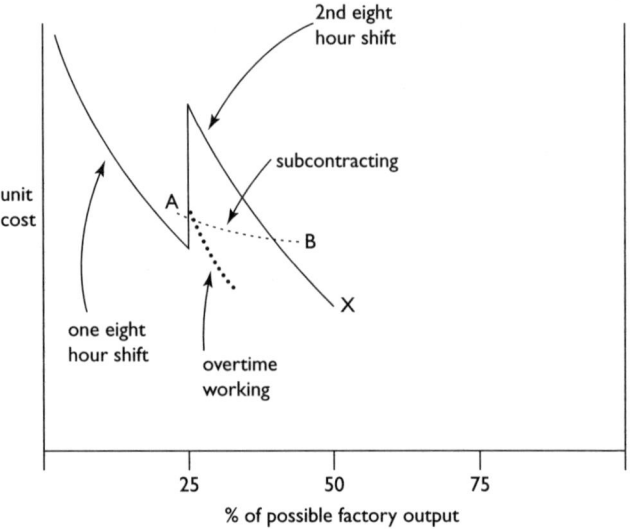

Figure 4.7 Unit costs as a second shift, overtime or subcontracting is introduced

The costs of the alternative strategy of introducing a second shift are also shown on Fig. 4.7. The extra variable costs will be quite significant as considerably more indirect workers are required compared with overtime working. This causes an increase in unit costs until production levels approach the full two-shift capacity. Eventually economies of scale once again prevail and the unit cost falls below that achievable by extended overtime working at some point such as point X on Fig. 4.7. Eventually it will fall below that at 24 per cent production and go on to a new minimum at 48 per cent production output.

The process now repeats itself as a third shift or overtime is introduced. However if three week-day shifts are in operation only weekend overtime can take production over 72 per cent as a fourth shift cannot be introduced. This is shown in Fig. 4.8.

Another typical problem occurs when demand falls below that which will occupy full shift levels (24 per cent or 48 per cent or 72 per cent). The possibility of subcontracting out the manufacture or service to another provider may become attractive. The unit costs should fall on lines such as A–B in Fig. 4.7 for production levels below 24 per cent. At very low production levels the subcontracted parts ought to be cheaper as they should be made under higher production level conditions elsewhere. This differential should decrease as production levels reach full shift capacity and finally they should be more expensive due to the fellow producer's profit margin. However, there are many problems with subcontracting as much information often has to be made available to the competitor. For example product specifications, suppliers, mailing lists and so on. Hence careful consideration by a value engineering team should be made before such decisions are reached.

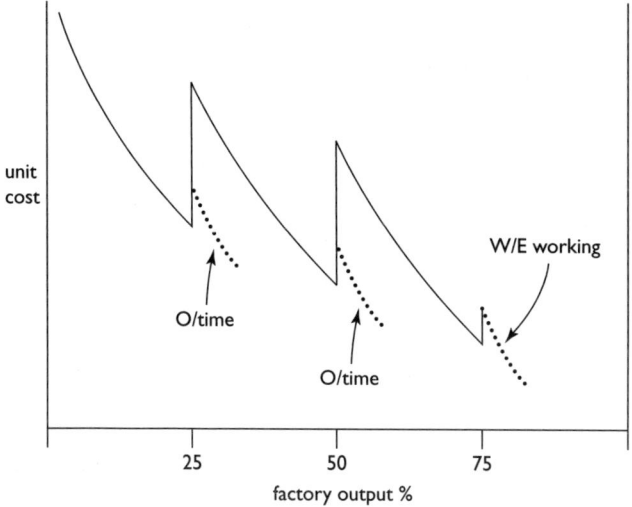

Figure 4.8 Unit costs for three shift + weekend working

At the other extreme there is the problem of achieving the maximum possible output. One common approach to achieve 100 per cent production levels is to run two 12 hour shifts on a four

64 Optimizing production rates and working practices

days on four days off basis. Four teams of workers are needed to operate such systems. These working practices have to be adopted regardless of unit costs where continuous production is more important than the actual amount of production. Public utilities, such as power and water supply usually operate in this fashion, at least for the fraction of the direct workforce involved in supply. Such supplies must be provided round-the-clock.

Two 12 hour shifts can cope with fluctuating production levels right up to 100 per cent capability if required as shown in Fig. 4.9. Production costs are also compared with a 3 × 8 hour shift pattern.

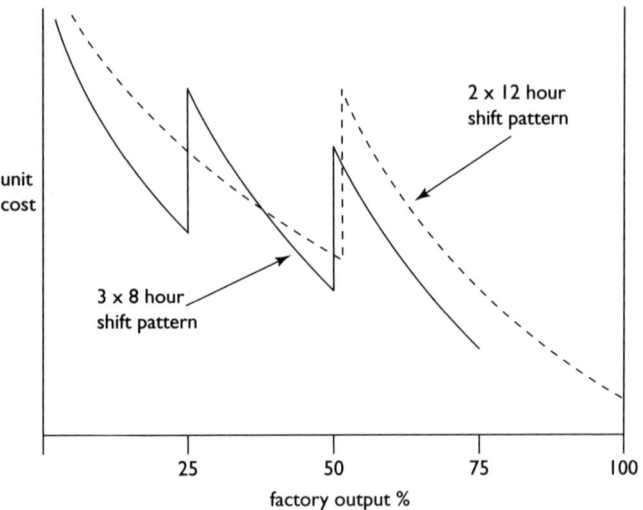

Figure 4.9 Unit costs as a function of output level for 2 × 12 hour shifts.

As can be seen the lowest unit cost is not achieved until 100 per cent output. Hence there are several production levels where the choice between overtime and the introduction of different shift patterns is difficult. The choice is often between an increasingly jaded workforce due to extensive overtime working or the trauma of redundancy if an additional shift is introduced which has later to be withdrawn. Hence careful considerations are needed when additional work is accepted. Such situations may require the VA team to be reconstituted to take an overall view of the situation (back to Chapter 1!) Here the principles can be used to assess the benefit of particular production levels in a multi-discipline way.

Objectives

After studying this chapter you should be able to:

1. Describe how it is not always beneficial to operate production and service systems at their highest possible output rates.
2. Explain why care has to be taken when projects are undertaken in terms of tooling and specialist equipment charges.
3. Explain why the choice between overtime working and the introduction of an additional shift needs careful consideration.
4. Explain why it may be sensible to subcontract manufacturing and service requirements.
5. Describe why a VA team might be needed to explore production level policy.

5 Displaying data and calculating averages

Any project will involve the collection and analysis of data. This may occur before, during and after the deliberations of the project team. Hence it is important to give an introductory overview in the next three chapters of the sorts of analysis that can be handled by the team. However if a project depends on a crucial statistical outcome it is better to include a statistician in the team!

Aims

- To consider how data is best presented in the particular context.
- To show how graphical representations are linked with analytical equations
- To show how logarithms and exponential relationships can be manipulated
- To show the advantages and disadvantages of popular ways of estimating 'average' values.

Collecting and displaying data

It is important to collect meaningful data during the course of a project. This will often mean that the project team must do considerable initial work to establish the important variables. It must also ensure that the data requested is sound and is not corrupted by unforeseen effects.

For instance, during the 'seventies' it appeared that French cigarettes were 'healthier' because rates of death from lung cancer and heart disease were lower in France than many other comparable

Displaying data and calculating averages 67

countries. However, the data failed to recognize the very strong link between drinking and smoking. As will be shown later, France has a very high death rate from alcoholism and cirrhosis of the liver and quite simply many French smokers had already died of alcohol-related illness before they had a chance to develop lung cancer. Hence, the simplistic conclusions drawn were not particularly useful to anyone other than the manufacturers of French cigarettes.

When collected, the presentation of data also requires considerable thought. Common methods are tabulated data, pie charts, histograms and graphical correlations between two or sometimes three variables.

Tabulated data

For display purposes, especially to people who are seeing the data for the first time, tables must be as simple and clear as possible. Table 5.1 was presented during December 1989 in the Guardian newspaper and is fairly complex. It is probably fine for someone with time to study it and the opportunity to return for further investigation. However, if it had been presented by a project team to an evening meeting at a political club to encourage/warn local members, it would probably have been too complex. Shown in Table 5.2 is a selection of the data, presented in a more suitable way for that occasion.

Notice the following points:

1. The data has been presented to show the change during the two months.
2. The SDP and L.Dem. results have been combined following the lead shown by NOP in November.
3. The percentages have been totalled to reflect the rounding errors, accuracy of the figures and number of spoilt returns (and to prevent unnecessary and irrelevant questions at the meeting about arithmetic!).
4. The averages have been weighted to take the different sample sizes into account.
5. Only polls available on both months have been used.
6. A final row showing the month's swing has been included.

68 Displaying data and calculating averages

Table 5.1

Fieldwork	Poll	Sample	Con.	Lab.	L.Dem.	SDP	Grn.
10–11.11	ICM/*Guardian*	1416	36	49	6	3	3
8–13.11	NOP	1628	34	46	9	†	6
22–23.11	Harris/*Obsvr*	1037	36	47	9	1	4
23–24.11	MORI/*S.Times*	1068	37	51	4	3	4
24–25.11	ICM/*S.Corr*	1460	38	48	3	4	5
24–26.11	ASL*	801	37	44	4	3	8
29.11–4.12	NOP	1760	38	45	7		5
1–4.12	Gallup/*D.Tele*	950	37.5	43.5	9	4	
8–9.12	ICM/*Guardian*	1333	37	49	4	3	4
12–14.12	Harris/*Obsvr*	978	39	46	6	1	5
Guardian average of last 5			38	46	6	5	3
General election June 1987			43	32	Alliance 23		

Where two or more polls were sampling at the same time, only one average is given. *Telephone poll. NOP no longer records a separate figure for the SDP but includes them in L.Dem.; average figures are adjusted accordingly.

Table 5.2

NOVEMBER FIGURES

Poll	Number of people	Percentages Con.	Lab.	L.Dem.	Others	Totals
NOP	1628	34	46	9	6	95%
ICM/*Guardian*	1416	36	49	9	3	97%
Harris/*Obsrvr*	1037	36	47	10	4	97%
Weighted Average	4081	35%	47%	9%	4%	95%

DECEMBER FIGURES

Poll	Number of people	Percentages Con.	Lab.	L.Dem.	Others	Totals
NOP	1760	38	45	12	0	95%
ICM/*Guardian*	1333	37	49	7	4	97%
Harris/*Obsrvr*	978	39	46	7	5	97%
Weighted Average	4071	36%	47%	9%	3%	95%
CHANGE		+1%	0%	0%	–1%	

Pie charts

Pie charts are useful in order to display information about proportions of comparable data in an easy way to non-experts. Table 5.3 gives the weekday sales by volume of drink in a particular public house. A pie chart may be constructed by making the 360° of the whole pie equal to the total volume sold and each slice a portion of this where the angle at the apex of the slice represents the relative proportion.

Hence for Beers the angle is given by $120/265.5 \times 360° = 168°$ (to the nearest degree) and the rest of the pie chart looks as shown in Figure 5.1

Table 5.3

Drink category	Volume (litres)
Beers (inc. shandy)	120
Lagers	75
Soft drinks (inc. hot beverages)	50
Spirits	1.5
Cider	10
Total	256.5 litres

However this does not give much useful information as a volume basis is hardly a fair comparison of the relative importances of the different drinks in trade terms.

The sales price per litre could be used to modify the data and in relative terms the daily sales would be as given in Table 5.4. Notice that to avoid inflation making prices look quaint they are given relative to beer which is given the value 1 unit per litre. The pie chart now displays quite a different picture as shown in Fig. 5.2.

This still does not tell the whole commercial story. The relative income from the different classes of drink reveals a different spread of data. This is because profit proportions are different on different products. These figures are very difficult to obtain from the industry but comparisons with prices charged in other outlets reveals the data shown in Table 5.5.

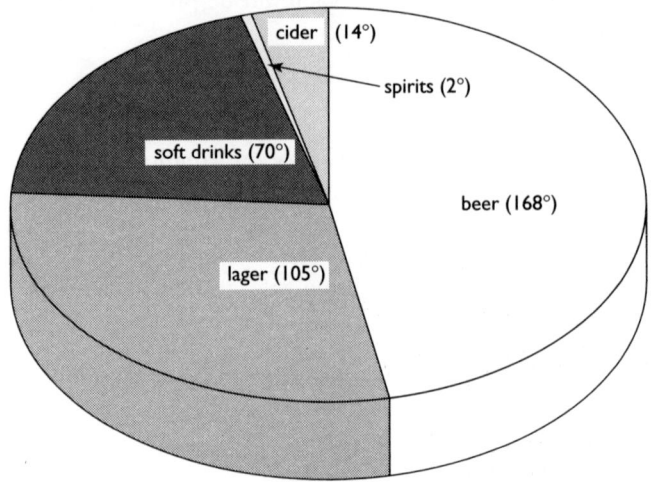

Fig. 5.1 Pie chart of drinks sales by volume

Table 5.4

Drink category	Daily volume	Relative price per litre	Relative daily sales value
Beer	120	1.0	120
Lager	75	1.2	90
Soft drinks	50	7.4	70
Spirits	1.5	70.0	105
Cider	10	1.25	12.5
		Total	397.5

The pie chart for the relative gross incomes from the different categories of drinks is shown in Fig. 5.3.

These three different pie charts all give instant images of the situation. First, large quantities of beer and lager; second, similar takings from beer, lager and spirits and, finally, most relative gross income from spirit sales. However, the whole commercial situation is much more complex than this. You would soon go out of business running a 'spirits-only' pub. Breweries also have many agree-

Displaying data and calculating averages 71

ments with landlords that influence the situation. The main point is that pie charts can give easily recognized summaries of certain types of information.

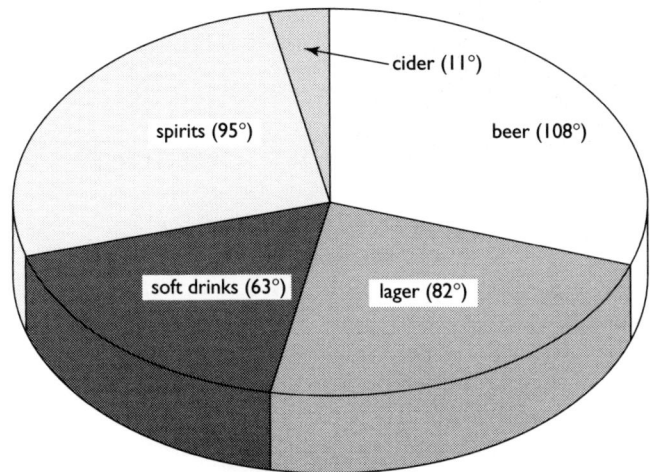

Fig. 5.2 Pie chart for drinks sales by relative daily value

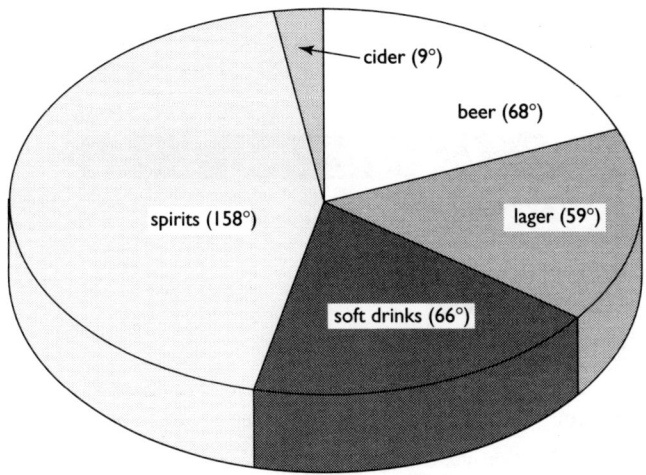

Fig. 5.3 Pie chart for drinks sales by relative gross income

Table 5.5

Drink	Daily relative sales value	Proportion of profit	Relative gross income
Beer	120	30%	36
Lager	90	35%	31.5
Soft drinks	70	50%	35
Spirits	105	80%	84
Cider	12.5	40%	5
		TOTAL	191.5

Histograms and bar charts

Although pie charts have much visual appeal, the relative amounts of the portions shown are related to the angles involved and people are not experienced at making numerical comparisons in this way. They know which slice of a pie looks the bigger but can't tell you by how much!

Bar charts display similar information but here the length of the bar conveys the important data and semi-quantitative comparisons are easier. The width of the bar usually has little or no significance. Fig. 5.4 shows the relative gross income from the drinks categories of Fig. 5.3.

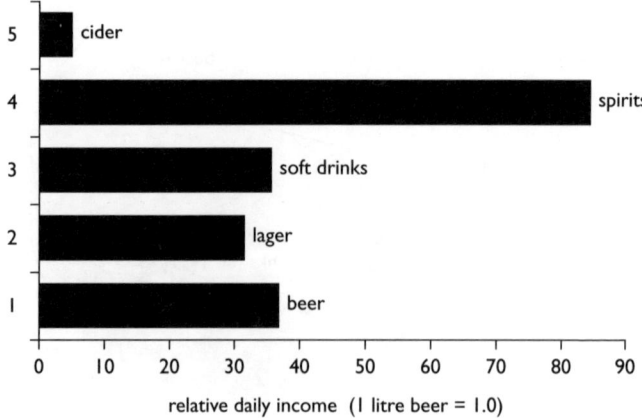

Figure 5.4 Bar chart of drinks income from Fig. 5.3

Displaying data and calculating averages 73

Histograms are similar representations. However the width of the bar symbols is usually also significant.

Shown in table 5.6 are the results of a traffic survey which counted the number of heavy goods vehicles per hour along a narrow country lane that was a short cut between two motorway class roads. The results were accumulated by shifts of observers in a continuous 400 hour period.

Table 5.6 HGVs per hour for 400 hours

2	2	4	4	4	5	2	4	7	7	4	7	5	2	8	6	7	4	3	4
3	3	2	4	2	5	4	2	8	6	3	6	6	10	8	3	5	6	4	4
7	9	5	2	7	4	4	2	4	4	4	3	5	6	5	4	1	4	2	6
4	0	4	7	3	2	3	5	8	2	9	5	3	9	5	5	2	4	3	4
4	0	5	9	3	4	4	6	6	5	4	6	5	5	4	3	5	9	6	4
4	4	5	10	4	4	3	8	3	2	0	4	0	5	6	4	2	3	3	3
3	7	4	5	0	8	5	7	9	5	8	9	5	6	6	4	3	7	4	4
7	5	6	3	6	7	4	5	8	6	3	3	4	3	7	4	4	4	5	3
8	10	6	3	3	6	5	2	5	3	11	3	7	4	7	3	5	5	3	4
0	3	7	2	5	5	5	3	3	4	6	5	6	1	6	4	4	4	6	4
4	2	5	4	8	6	3	4	6	5	2	6	6	1	2	2	2	5	2	2
5	9	3	5	6	5	6	5	7	0	3	6	5	4	2	8	9	5	4	3
2	2	11	4	6	6	4	6	2	5	3	5	7	2	6	5	5	1	2	7
5	12	5	8	2	4	2	0	6	4	5	0	2	9	1	3	4	7	3	6
5	6	5	4	4	5	2	7	6	2	7	3	5	4	4	5	4	7	5	4
8	4	6	6	5	3	3	5	7	4	5	5	5	6	10	2	3	8	3	5
6	6	4	2	6	6	7	5	4	5	8	6	7	6	4	2	6	1	1	4
7	2	5	7	4	6	4	5	1	5	10	8	7	5	4	6	4	4	7	5
4	3	0	6	2	5	3	3	3	7	4	3	7	8	4	7	3	1	4	4
7	6	7	2	4	5	1	3	12	4	2	2	8	7	6	7	6	3	5	4

In this form the data looks very difficult to comprehend. However it is exactly the right kind of data that is best shown by histogram. The first step is to reassess the data by dividing it into different classes. Table 5.7 shows the results of counting the number of observations that occur with particular numbers of HGVs per hour from zero to twelve in units of one.

This can now be plotted as a histogram as shown in Fig. 5.5. In this form the data is much more understandable. At a glance it can be seen that the most common rate is four vehicles per hour and the

74 Displaying data and calculating averages

maximum rate is 12 vehicles per hour. For 10 hours out of 400 there was a zero HGV traffic rate. The horizontal axis has been arranged with the bars positioned symmetrically about the hourly HGV rates.

Table 5.7 HGVs per hour classified by number of observations

HGVs per hour	0	1	2	3	4	5	6	7	8	9	10	11	12
No. of observations	10	10	43	53	86	70	54	37	18	10	5	2	2

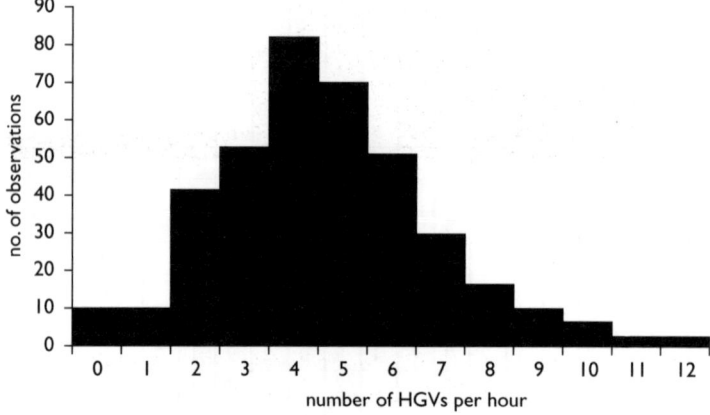

Figure 5.5 Histogram of HGV traffic flow

The frequency of particular traffic flow rates is represented by the heights of the bars and this can be related to the areas of the bars if the class intervals, represented by the widths of the bars, are equal.

If the infrequent data from the very low and very high flow rates was collected in bigger class intervals as shown in Table 5.8. It would then be necessary to adjust the heights of the histogram bars to maintain this area relationship. This is shown in Fig. 5.6 where a 3 unit wide bar at 21 on the vertical axis represents the 63 hours when traffic flow rates were below 3 vehicles per hour. A similar modification has been made at the high flow rates.

Displaying data and calculating averages

Table 5.8 HGV flow rate – frequencies with large class intervals at low and high values

HGVs per hour	0, 1 or 2	3	4	5	6	7	8 to 12
Number of observations	63	53	86	70	54	37	37

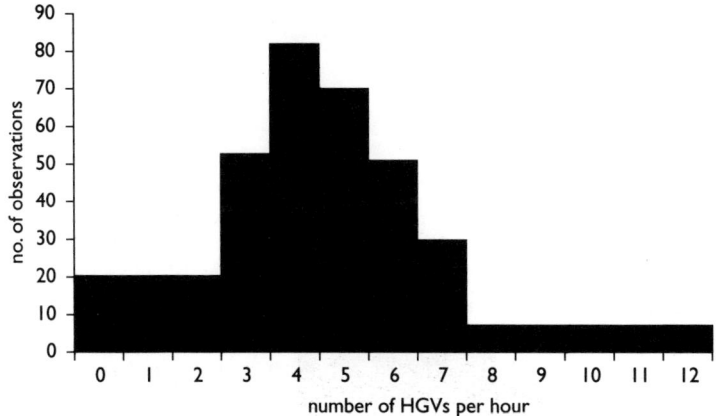

Figure 5.6 HGV traffic flow histogram with modified class intervals

The technique of analysing data through the frequency of particular values to produce a frequency distribution has much deeper statistical significance than just displaying the data and such distributions will be returned to in Chapters 6 and 7.

Graphical correlations

Two variables
The way one observation depends on another is usually best presented in the form of a graph. It is conventional to plot the independent variable on the horizontal axis and the dependent variable on the vertical axis. The independent variable is what can be changed in a pre-determined fashion and the dependent variable is that feature which changes in response. For instance in fluid mechanics the pressure on particular fluid can be easily changed to influence the flow rate in a pipe of certain dimensions. Here the pressure is the independent variable.

76 Displaying data and calculating averages

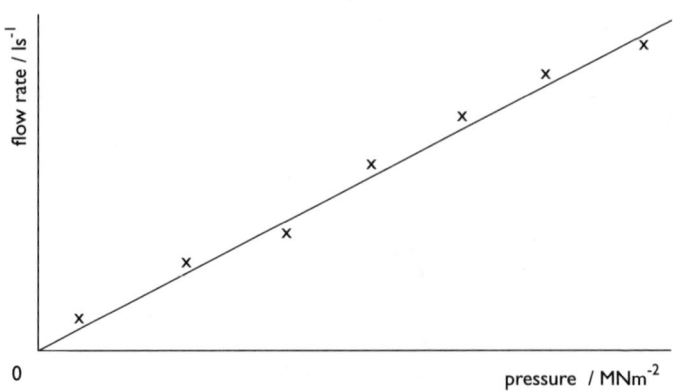

Figure 5.7 How the flow rate (dependent variable) is changed by pressure (independent variable)

Fig. 5.7 shows the results for a Newtonian fluid that shows a linear (straight line) relationship between pressure and flow rate. The relationship is determined by the viscosity of the fluid. Here the points are close to the line because such a simple experiment can be conducted under closely controlled conditions. It can be said that there is a **strong correlation** in the data.

Similar graphs are still useful even when the correlation is not so strong. Fig. 5.8 shows the relationship between alcohol consumption per person per year (average) and death rates per 100 000 population due to alcohol related illness, for various countries as shown in Table 5.9. Here another technique has been used to add another 'dimension' to the graph by using letters to label the points to identify the country. Alcohol consumption has been made the independent variable in order to show how death rates depend on it.

Once again you can see the general trend and the use of the letters enables you to see how different countries compare at a glance.

Table 5.9 Average alcohol consumption and death rate. Extracted from J.F. Osborn. *Statistical Methods in Medical Research (1979)* Blackwell Scientific.

Country		Alcohol consumption (litres/person/year)	Cirrhosis and alcoholism (death rate/100 000)
France	(F)	24.7	46.1
Italy	(I)	15.2	23.6
Germany	(G)	12.3	23.7
Austria	(A)	10.9	7.0
Belgium	(B)	10.8	12.3
USA	(U)	9.9	14.2
Canada	(C)	8.3	7.4
England	(E)	7.2	3.0
Sweden	(S)	6.6	7.2
Japan	(J)	5.8	10.6
Netherlands	(H)	5.7	3.7
Norway	(N)	4.2	4.3

Three variables

Fig 5.8 begins to address this issue. However when a quantitative third variable is considered it is not possible to allow it to change in a continuous fashion without resorting to three dimensional representations where lines become planes.

Fig 5.9 shows what is known as the p–V–T surface that describes the relationship when the pressure, volume and temperature are changed on a fixed amount of a gas. The relationship between the three creates the curved planar surface as shown. However such curved places are not easy to interpret and take readings from. An alternative is to allow the third variable to have specific values and plot a series of straight lines on a graph.

Fig. 5.10 shows a plot for the Newtonian fluid flow experiment conducted at three different temperatures.

In this form readings can be easily extracted from the graph and the message here is that high temperatures make the liquid less viscous and higher flow rates are observed for the same applied pressure. However it was necessary to fix one variable (temperature) to certain specific values before the experiment started in order to plot the data in this way.

78 Displaying data and calculating averages

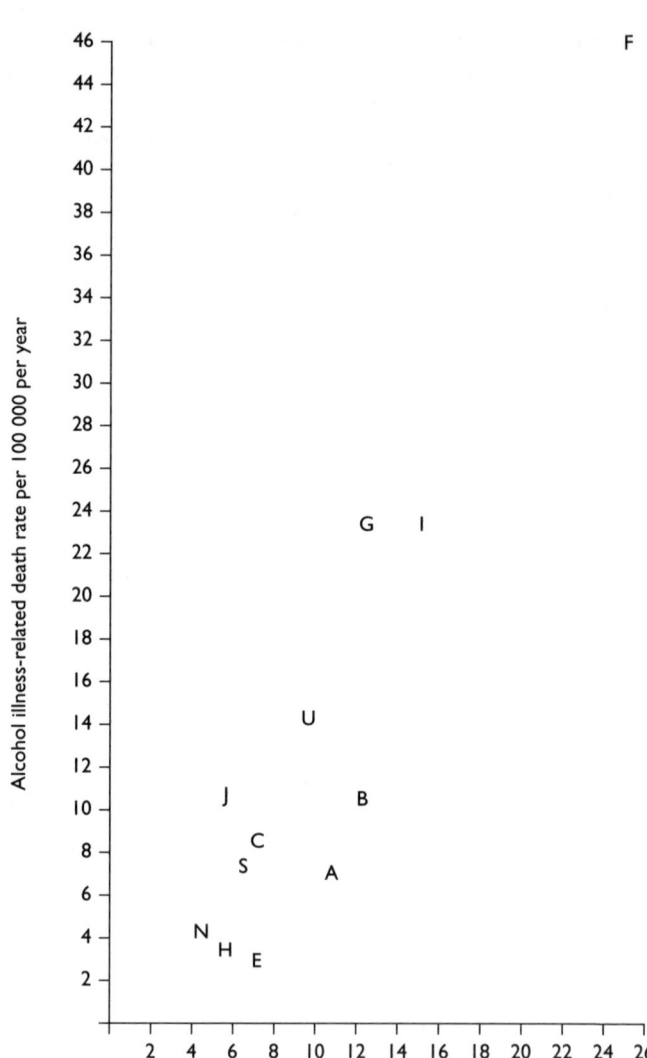

Figure 5.8 Death rates and drinking in different countries

Displaying data and calculating averages 79

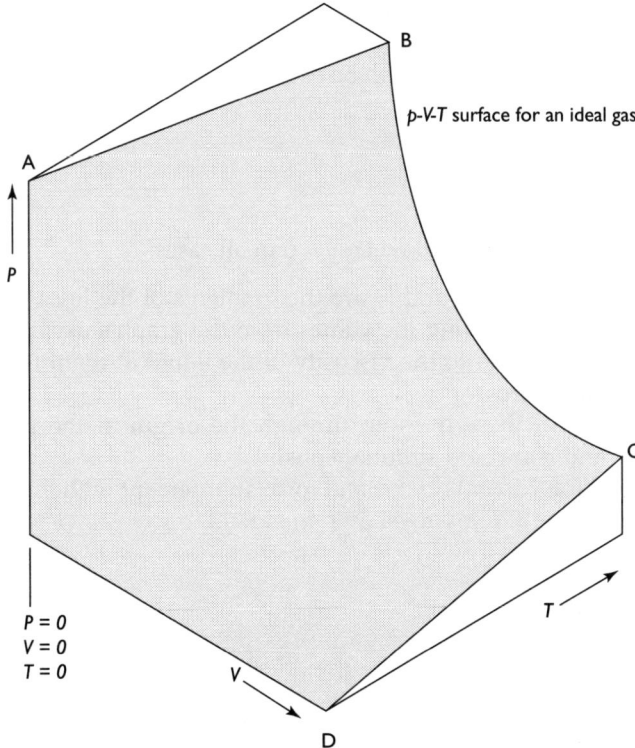

Figure 5.9 Three variables as a plane

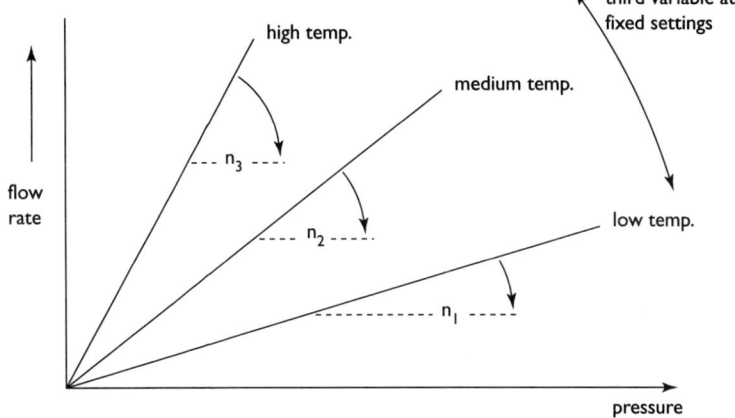

Figure 5.10 Three variables as a family of lines

Equations of lines on graphs

The three lines in Fig. 5.10 can be expressed as equations

$$y = n_1 x$$
$$y = n_2 x$$
$$y = n_3 x$$

where $n_3 > n_2 > n_1$ and $x = 0$ when $y = 0$ in all cases.

The constants n_1, n_2 and n_3 are the gradients of the lines which can be found by reading the values from the graph's axes. These gradients are related to the viscosity of the liquid at the three different temperatures.

Note that all these lines go through the **origin** of the graphs where x and y are zero simultaneously.

Many graphs do not do this and form an **intercept** with the axes. Two such straight lines are shown in Fig. 5.11.

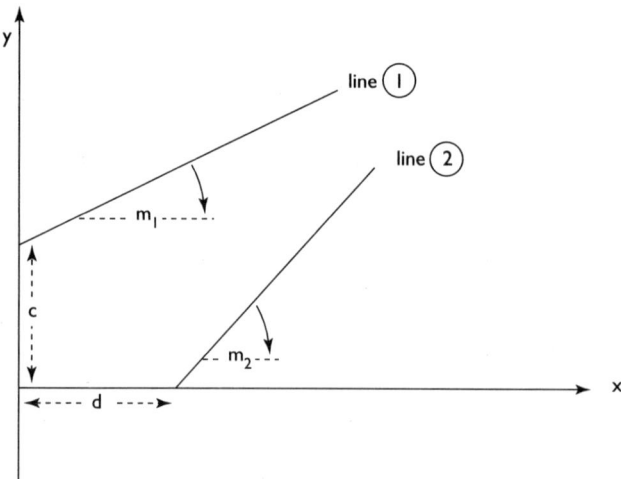

Figure 5.11 Linear relationships with intercepts on the axes

The two relevant equations are:

line 1 $y = m_1 x + c$ (when $x = 0$, $y = c$)

line 2 $x = \dfrac{1}{m_2} y + d$ (when $y = 0$, $x = d$)

If these were sales figures (y) for two different offices, plotted against time (x) it can be seen that the lines will eventually cross at a particular value of x. This can be found by drawing the graphs and **extrapolating** the data to find where the lines cross or it can be done by calculation.

At the cross-over (again called an intercept) both values of x are equal and so are the two values of y.

Hence

$y = m_1 x + c$ (line 1)

$y = m_2(x - d)$ (rearranged version of line 2)

since the y values are equal

$m_1 x + c = m_2(x - d)$

$\therefore m_2 x - m_1 x = c + m_2 d$

$\therefore x(m_2 - m_1) = c + m_2 d$

$\therefore x = \dfrac{c + m_2 d}{m_2 - m_1}$

Hence the time when the offices have identical sales figures can be predicted. The sales level in both offices at that time can also be calculated by substituting this value of x back into either of the original equations.

Manipulating power law and exponential relationships

Many lines, even curved ones, on graphs can be given equations and some common ones are shown in Figure 5.12.

The origin of such equations is not particularly relevant to projects. However it is useful to be able to manipulate such data for display and analysis purposes.

In project work the parabola and exponential curves are very common. For instance, recording sales in the music business often follow the latter type of curve as artists get well-known and promoted. There is a relatively slow start and then sales go up through the roof! (So do their fees!!)

82 Displaying data and calculating averages

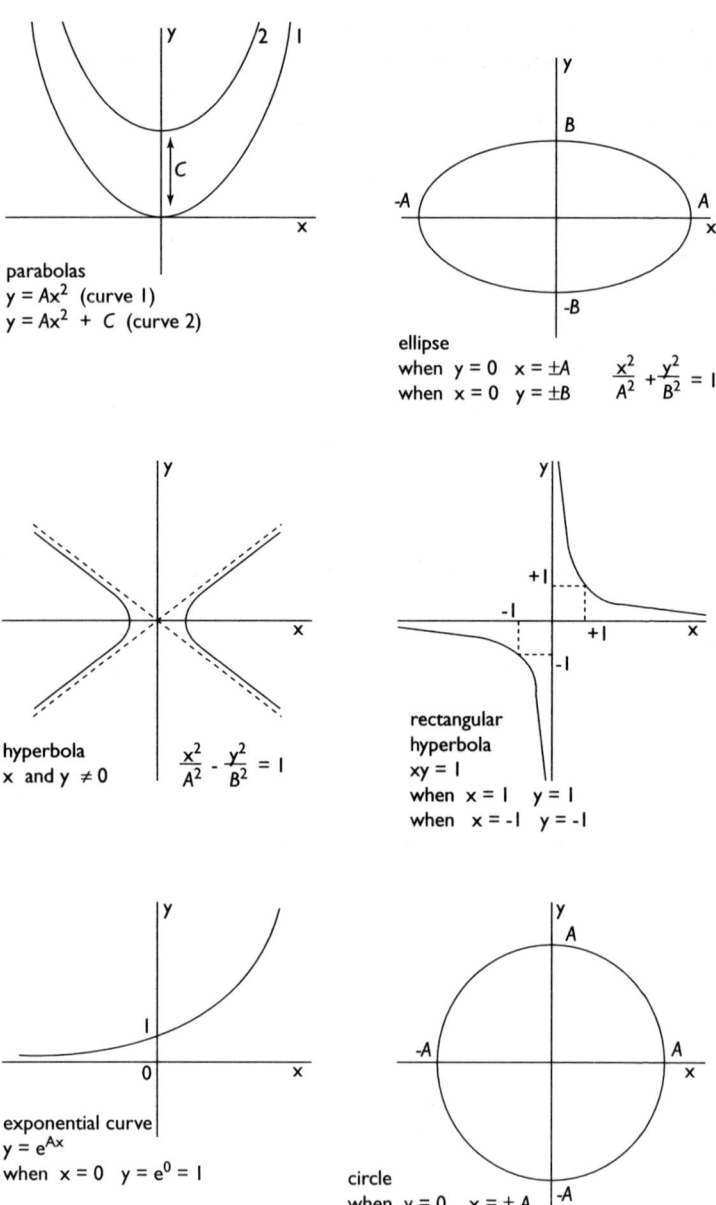

Figure 5.12 A selection of curves and their equations

This can cause problems for graphical representation as extremely large numbers have to be shown alongside extremely small numbers. This problem can be solved by taking logarithms of the equation. This also has the benefit of turning the data into a straight line plot.

For instance $y = Ax^2$ (parabola 1 on Fig. 5.12)

becomes $\log y = \log A + 2\log x$

Now a plot of log y versus log x gives a straight line of gradient = 2 and intercept log A on the log y axis as shown in Fig. 5.13. This occurs when $\log x = 0$.

Conversely if you don't know the equation but you do have x and y values, you can find the equation. If a log–log plot gives you a straight line with a gradient 'n' and an intercept 'B' on the log y axis, you can immediately state the equation as

$y = Bx^n$

This is often called a power-law relationship because y varies with some power of x.

The exponential relationship is so called because y varies as the number e (2.71828....) raised to the power Ax where A is a constant.

Hence $\quad y = e^{Ax}$

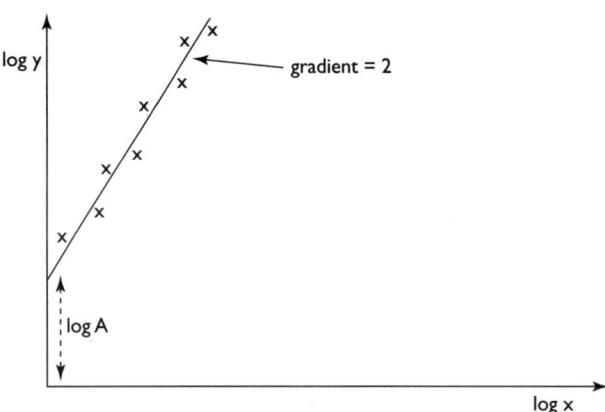

Figure 5.13 log-log plot of $y = Ax^2$

Now a semi-logarithmic plot of $\log_e y$ against x will give a straight line through the origin if the data follows an exponential relationship. The constant A can be found from the gradient by reading relevant values off the axes.

This is because $\log_e y = Ax$

Logarithms to the base e are known as natural logarithms and the symbol $\log_e$ is sometimes replaced by ln. Hence ln x is the same thing as $\log_e x$.

Fitting equations to unusually shaped graphs

Almost any curve on a graph can have an equation fitted to it. This is done using the technique of polynomial curve fitting. This can be done on the raw data utilizing standard software packages. It is beyond the scope of this text to present a background to this technique. However once the equation has been established it can be used to interpolate data within the range covered. Extrapolation is more difficult and should be approached with caution.

Measures of central tendency

People like to deal with typical values so we often call these averages. When data is collected to give individual data points that are not related to each other and do not influence each other then the results can usually be described as having a tendency towards a particular 'central' value. However, there are several ways of estimating this central value. The average weight of a group of adults is a typical task that could be addressed in this way.

Arithmetic mean ($\bar{x}_a$)

In this case the average weight would be correctly termed the *Arithmetic mean* and if our values are $x_1, x_2, x_3.......x_n$ then their arithmetic mean is given the symbol ($\bar{x}_a$). It is determined by adding all the values together and dividing by the number in the group. This summation is given the symbol Σ (Greek letter sigma, upper case).

Therefore:

$\sum_{i=1}^{n} x_i$ means add all the x values from x_1 up to x_n

hence $\bar{x}_a = \dfrac{\sum_{i=1}^{n} x_i}{n} = \dfrac{x_1 + x_2 + x_3 \ldots x_n}{n}$

this is often shortened to $\bar{x}_a = \dfrac{\Sigma x}{n}$ and means exactly the same.

There are some useful things about arithmetic means

1. First if you multiply the arithmetic mean by the numbers of readings taken you get the sum Σx. So if you happened to know the average height of people in a country (h) and the population (N) the distance needed to lay them out end to end would be $N(h) = \Sigma h$

   In more practical terms if you knew that the arithmetic mean of the weight of flour in a bread loaf was 300 grams and that 5 million loaves were sold daily in a country, the national daily flour requirement for bread alone would be 1500 million grams, which is 1500 tonnes.

2. Second, if you sum the differences between the individual values and the arithmetic mean value, they cancel themselves out to give zero. In symbols this is written as

$$\sum_{i=1}^{n}(x_i - \bar{x}_a) = 0$$

3. Finally, the origin of the data points can be adjusted by adding or subtracting any value A from each data point. The scale or range of the data can be adjusted by dividing each data point by any positive non-zero number (C). Hence each new data point u_n is given by

$$u_n = \dfrac{x_n - A}{C}$$

To decrease the origin, A should be a positive number. To increase the origin, A should be a negative number. To reduce the scale, C should be greater than one. To increase the scale, C should be less than one. The new arithmetic mean is $\bar{U}$ and the original is $\bar{X}$.

where $\bar{X} = C\bar{U} + A$

86 Displaying data and calculating averages

Airliner operators might use average national weights in their calculations to estimate how much extra freight could be added to a 747 full of Japanese compared with Americans. This could be done by having different 'C' factors for each nationality. They might also be able to shift the origin of average weight data by using the 'A' factor to compensate for the different luggage requirements for a business charter against a heavily-laden skier's charter flight. Regardless of their size, skier's extra luggage usually weighs about 15 kg per person.

Mid-range ralue (x_{mr})

The range of a set of data is simply the maximum minus the minimum value and hence the mid-range value is simply half this value added to the minimum value.

Hence the range = $x_{max} - x_{min}$

and $x_{mid\text{-}range} = x_{min} + \dfrac{(x_{max} - x_{min})}{2}$

This value might be used as a very rough estimate of the mean if it is not worthwhile calculating the arithmetic mean. For instance airline luggage manufacturers might consider the range of luggage weights associated with airline passengers and decide to pitch their standard suitcase at mid-range weight capability. It would not be worthwhile monitoring large numbers of passengers because the exact arithmetic mean of their luggage weights would not be much more revealing.

Median (x_m)

The median of a set of data points is usually defined as the middle value when they are all placed in ascending order. This is easy for sets containing an odd number of values. For even numbers the arithmetic mean of the middle two is found to give the median value. This concept is analytically weak and not as useful as the arithmetic mean. However its value comes from its insensitivity to the presence of a few extreme values.

For instance if you were to develop a new product you might be interested in calculating the average disposable income of a group of citizens. However, you should be aware that the arithmetic mean in many communities would be affected by the presence of a few very rich locals. Hence if the data were arranged to establish the median, the incomes of the few very rich people would not distort the data and give an over-estimate of the typical disposable income.

Mode (x_r)

The mode or modal value of a set of ungrouped data is the value that occurs most frequently. It is of most use when the data is grouped into particular ranges. For instance if you asked men at random what shoe size they take the most frequent response in Britain would probably be 8. Hence the mode of your sample would be 8.

Hence, if you were a market trader and were offered a 'bulk buy' of men's shoes of one particular size you would recall that the modal value for shoe sizes is 8 and choose it. The arithmetic mean of all men's shoe sizes might be 7.8 but people don't make shoes of this exact size. Shoe manufacturers go one stage further and acquire data that gives the frequency of responses in all size ranges and produce shoes at a corresponding frequency distribution. This requires actual sizes of feet to be grouped into size categories and the magnitude of each category found. This grouping of data will be considered later in this chapter.

Geometric mean ($\bar{x}_g$)

This is an analytically useful measure of central tendency and may be compared to the arithmetic mean. It is defined as the nth root of the product of the individual values in the group of data

Hence $\bar{x}_g = (x_1 \bullet x_2 \bullet x_3 ..... x_n)^{\frac{1}{n}}$

For example the data group

4.1, 4.2, 4.3, 4.4, 4.5, 4.6, 4.7, 4.8, 4.9

has a geometric mean given by

$\bar{x}_g = (4.1 \times 4.2 \times 4.3 \times 4.4 \times 4.5 \times 4.6 \times 4.7 \times 4.8 \times 4.9)^{\frac{1}{9}}$

$= (745520.86)^{0.1111}$

$= 4.492$

comparing this with the arithmetic mean

$$\bar{x}_a = \frac{4.1 + 4.2 + 4.3 + 4.4 + 4.5 + 4.6 + 4.7 + 4.8 + 4.9}{9}$$

$\bar{x}_a = 4.500$

Hence the geometric mean is smaller than the arithmetic mean.

Now if $\bar{x}_g = (x_1 \bullet x_2 \bullet x_3 .... x_n)^{\frac{1}{n}}$

$\log \bar{x}_g = \frac{1}{n} \log(x_1 \bullet x_2 \bullet x_3 .... x_n)$

$= \dfrac{\log x_1 + \log x_2 + \log x_3 ....+ \log x_n}{n}$

this may be stated in general terms as

$\log \bar{x}_g = \dfrac{1}{n} \Sigma \log x_i$

This implies that the logarithm of the geometric mean is equal to the arithmetic mean of the logarithm of individual values in the data set.

N.B. Negative and zero values in the data set make it analytically difficult to attach significance to the geometric mean and this estimate should be avoided in such circumstances.

Grouped data

For certain projects, data is collected in groups. This enables histograms to be plotted with each bar of the histogram representing the amount of data within that division. For example in Fig. 5.14 the response times of a hypothetical ambulance service have been grouped and plotted against the frequency of their occurrence. At the time of writing the UK target is to get all ambulances to their destinations in under 19 minutes in response to emergency calls. Clearly this data includes non-emergency call-outs.

Displaying data and calculating averages

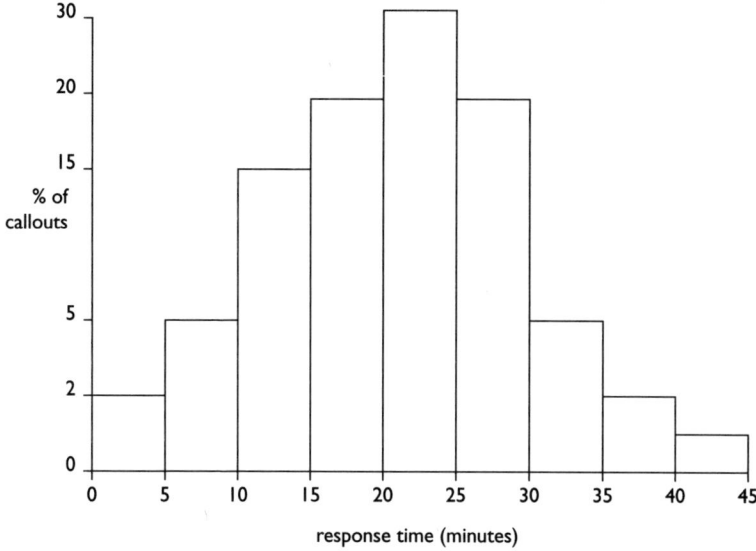

Figure 5.14 Ambulance response times plotted in 5 minute intervals and shown on a histogram

Whilst the data remains in these bands there is a limit to what can be calculated from it. However, for this set of data we can say:

i. The mid-range value is 22.5 minutes which must occur in the 20–25 minutes band.
ii. The median value occurs when adding from one end towards the other, the accumulated percentage of call-outs reaches 50 per cent. Clearly this also occurs in the 20-25 minutes band.
iii. The modal value, the most frequent callout band is also the 20–25 minutes range.

In order to calculate the arithmetic mean we must replace the bands with the arithmetic mean of their extreme values. This has been done in Fig. 5.15.

The arithmetic mean of all the data may now be found by taking the frequency of occurrence of each value into account and assuming that the arithmetic mean of the extreme values represents the arithmetic mean of all the data within the band. The process of taking frequency into account is often called 'weighting' the data and the weighted arithmetic mean x is given by

$$\bar{x} = \sum_{i=1}^{n} \frac{f_i x_i}{N}$$

where f_i represents the frequency of the ith band and x_i is the mean value of the ith band. N is the total number of data points in the set.

Hence for the data in Fig. 5.15, (assuming we have 100 call-outs) the calculations are shown in Table 5.10.

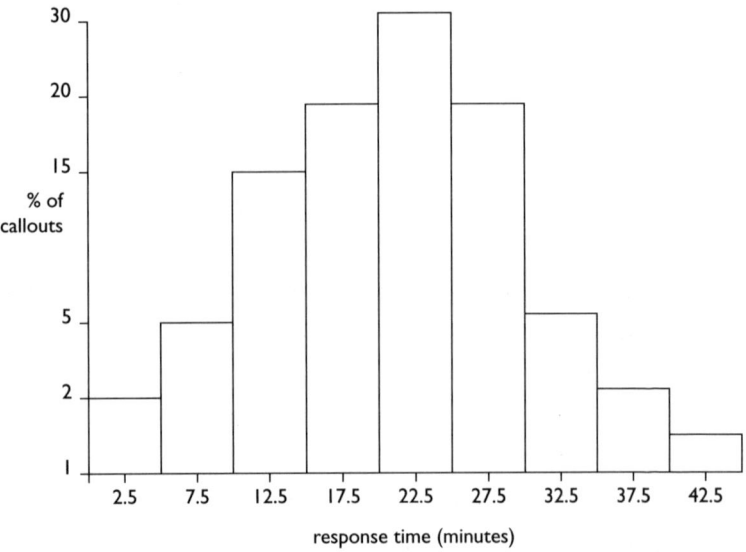

Figure 5.15 The extreme values of the bands in Figure 5.14 replaced by their arithmetic means

$$\therefore \bar{x} = \frac{2085.0}{100}$$

$= 20.85$ mins

Hence the average call-out time for this particular ambulance station may be calculated even though the data has been collected in groups, each of five minutes duration.

Table 5.10

x_i	f_i	$x_i f_i$
2.5	2	5.0
7.5	5	37.5
12.5	15	187.5
17.5	20	350.0
22.5	30	675.0
27.5	20	550.0
32.5	5	162.5
37.5	2	75.00
42.5	1	42.5
	100	$\sum f_i x_i = 2085.0$

Objectives

After studying this chapter you should be able to do the following.

1. Decide when information is best displayed as a table, graph, pie chart, bar chart or histogram.

2. Explain how an analytical equation may be related to the shape of a curve on a graph of two variables.

3. Explain how three variables may be represented on a graph.

4. Calculate the arithmetic mean for grouped and ungrouped data.

5. Recognize and define the following terms; geometric mean, mid-range value, media, mode, power law, exponential relationship, logarithmic axes.

6 Probability effects in projects

Aims

- To define the probability of an event
- To show how the probabilities of events may be combined
- To link probability effects with acceptable and unacceptable variations in quality
- To show how simple probability estimates can be used to assess project planning, risk assessment and control of quality.
- To show the origins of probability distributions

Probabilities of events

You cannot be certain of anything and as you will see, project outcomes reflect life and always carry uncertainties.

Hence the idea of the 'probability' of individual events occurring is a crucial concept to many situations. Much of the gambling industry is based upon individuals betting 'for' or 'against' probable events. Wagers are accepted at particular 'odds' that reflect the probability of the event occurring. These usually give small potential gains for likely events and high potential rewards for unlikely events. The odds are also slightly adjusted in the favour of the 'permanent' institution that accepts the bet. This may be a casino, bookmaker, lottery operator, etc. These are necessary to provide the infra-structure to maintain this sector of the entertainment industry. Illogical and ill-informed bets and secondary spending provide additional revenue.

Probability has significant importance in science and technology and we are surrounded by probability effects every day in our existence. The formal definition of the probability (p) of a particular event is as follows:-

$$p = \frac{\text{number of possibilities of the event occurring}}{\text{total number of possibilities}}$$

Acceptable variations in events

The arrival of an ambulance at a certain time after it has been requested has an associated probability. If the vehicle is requested from the station described in Fig. 5.14 you could convert the data in the graph into a series of probabilities.

Two times out of 100 it arrives in less than 5 minutes. This depends on your location and the traffic level at the time. Hence the probability is

$$p = \frac{2}{100} = 0.02$$.

The probability of it arriving in less than 25 minutes can be found by adding up the possibilities for all arrival times less than 25 minutes. Hence out of every 100 call-outs.

2 arrive in less than 5 mins
5 arrive in less than 10 mins. but more than 5 mins
15 arrive in less than 15 mins but more than 10 mins
20 arrive in less than 20 mins but more than 15 mins
30 arrive in less that 25 mins but more than 20 mins

Hence a total of 72 arrival times are less than 25 minutes and this event can be described by a probability where :-

$$p = \frac{72 \text{ arrivals}}{100 \text{ call-outs}} = 0.72$$

Notice how the probability is cumulative in terms of the arrival events less than our target.

Clearly the probability is related to the area under the histogram. Certainty, with $p = 1$ is represented by the whole area of the histogram bars. For the data provided you can be 'certain' that an ambulance will arrive in 45 minutes.

We usually have to be tolerant of such probability distributions and in general we accept certain variations in both products and services. We don't return 1 kilo of bread because it only weighs 990 grams. It might weigh 1010 grams tomorrow. We don't send back a consignment of planks because one is split. However, at some point, variations in products and services may become unacceptable.

Unacceptable variations

These events are usually linked with the word 'quality'.

There are variations in the 'quality' of everything we do. 'Quality' may be the range of accuracy to which our machine tools mass produce components. It may be the number of assembly defects that occur in a finished consumer durable (car, TV, washing machine, etc.). In the service sector it may be the response time of an ambulance or telephonist, or the number of trains that run to schedule per day. All these can be measured in quantitative terms.

There are variations in quality that are more difficult to measure. Hence these must be judged in qualitative terms. For example the performance of staff in banks, airports, hotels, etc. is certainly variable, but how do you measure it? It could be measured in response times or working rates, but that would not reflect the whole situation. You might issue questionnaires to customers and ask them to rate the service provided. Together these inputs might give a measure of quality that could be monitored throughout the year.

Why should we be concerned over variations in products and services? As already stated people and systems are tolerant of variability, but only up to a limit. It is useful to establish as early as possible if the variations are due to random or systematic fluctuations which are not likely to worsen. In this case the best course of action is not to interfere. However a more general decrease in quality may underlie these variations and this may be indicative of a long term deterioration in quality. For example Fig. 6.1 shows the change in ambulance response times for a particular service plotted each time an ambulance is called out over a period of a month.

Probability effects in projects

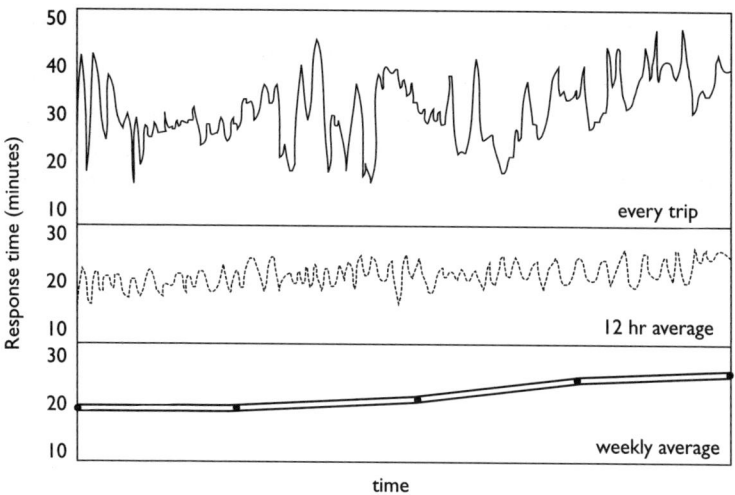

Figure 6.1 Responses times of ambulance call outs averaged out in three ways

The continuous line shows the variation in times for individual trips which includes changes in distance and traffic conditions. The dotted curve shows the same data where the averages are plotted for 12 hour periods to reveal a daily small increase in the afternoon and evening when traffic is the heaviest. The double line shows the data averaged each week to reveal a long term general increase. The recipients of the service may tolerate occasional response times above 25 minutes but at some point when the **average** response time is 25 minutes, especially for emergency calls, then intolerance of the service may occur and changes may be demanded to restore the average response time to below tolerable limits. The management team then needs to decide on whether to spend money to improve the service or to convince members of the public to reduce their expectations. The second recommendation might have resulted from a 'Value Analysis' which concluded that traffic levels and road systems combine to give negligible benefit from cash spent on improved equipment and communication. The VA team might advise that money is better spent on improving the medical capabilities of the ambulance crews in order to reduce the 'time to receipt of medical attention'. However this also means investing in extra ambulances and crews since the time taken by each call-out is increased.

For producers of engineering components similar decisions occur when equipment starts to wear or customer's specifications become more stringent. For example, a similar set of graphs would be obtained for the length of a small bolt machined on a machine tool. The variation of such a critical dimension could be affected by these factors.

(1) Random variations due to tool wear and inaccuracies in measurements and settings.

(2) Short term cyclic variations due to changes in temperature of the factory or machine tool.

(3) Long term irreversible changes due to general wear in the critical bearing surfaces within the machine itself.

What to do about variable quality

The examples chosen above represent two extremes of *acceptance sampling*. It is usually a statutory requirement to measure the response time of an ambulance for each and every trip. However for a small general purpose bolt, costing very little, checking the length of each one made would be economic madness! Hence it is necessary to adopt quality guidelines.

The consequences of 'out of specification' products or services reaching the customer fall into three broad possibilities.

1. The individual products may be returned for replacement or the particular service may have to be repeated.

2. If the fault levels reach an intolerable level the whole batch may be rejected or the service contract lost.

3. A *single* faulty product or service and the whole batch is rejected or the service contract lost.

For trading conditions that relate to the first category, the level of acceptance sampling and inspection can be determined by the relative costs involved.

If batch size = N, fraction defective = f (typical), cost of one item returned = C, inspection cost per item = I, then the cost of 100 per cent inspection = NI per batch, and the maximum cost of returns = fNC.

Now if the max. cost of returns is less than 100% inspection costs i.e. $fNC < NI$, it is probably not worth inspecting at all. But if 100% inspection costs less than the cost of a single return i.e. $NI < C$, then it is probably worth inspecting every item to ensure zero legitimate product returns or service repeats.

This inspection protocol is limited to a simple yes/no strategy. That means for example, measure the length of every bolt and record the time of every ambulance trip or not at all. However for many complex consumer durable products and service provisions it is the level of inspection between these two extremes that is important. For example you might consider running every washing machine you make and then clean out all traces of the process fluids so that it may be sold fully tested and spotlessly clean. Similarly you could pay an inspector to always follow your parcels delivery service to ensure that your service claims of 'time' and 'safe transit' are always achieved. However both these strategies would be economic madness.

Hence the SPOT CHECK, which means taking random samples to ensure that quality levels appear to be reaching acceptable levels is a commonly used approach. For example a supermarket chain might claim that queues at check-outs never to exceed 5 minutes but the manager is not going to sit there ensuring this is true.

In order to be able to justify such claims just enough staff are employed to appear to uphold it. In practice it may mean that for say 2 hours out of 50 hours, that queues exceed this criterion, even when all staff are working efficiently

Hence, a 'tolerable defect level' has been accepted. In this case it would be 2 in 50 or 0.04 expressed as a decimal fraction. Likewise manufacturers often have to accept tolerable defect levels in order that production rates are competitive. This means that batches of production are acceptable if they contain less than a critical level of defective units.

Clearly making a single spot check does not ensure that defect levels are being maintained at tolerable levels. A single secret inspection of a service provision also gives very limited information. A single component withdrawn from a batch for complete

98 Probability effects in projects

testing also gives minimal information about the batch defect level.

The key question is 'How many samples or spot checks need to be made to ensure that tolerable defect levels are being maintained?'

In order to answer this type of query we need to cover some aspects of probability theory.

Combining probabilities

If you toss a single perfect coin the chances of getting 'heads' is 50 per cent or 1 in 2 or 0.5 they all mean the same thing. However 0.5 is the most useful way to express the probability (p). The probability of getting a 'tails' is also 0.5.

This gives us our first rule:

The probabilities of all possible eventualities adds up to one.

In our example: p'heads' + p'tails' = 1

i.e. 0.5 + 0.5 = 1 (check)

If you toss two coins simultaneously the probability of getting two 'heads' is 0.5 × 0.5 = 0.25.

These events are linked because one event must be successful before it is worth considering the outcome of the second event.

This gives us our next rule:

The overall probability for two linked events is the multiple of their individual probabilities.

For two coins the probabilities of all the possibilities are as follows:

COIN A		COIN B		probability	
Heads	+	Heads	=	0.5 × 0.5	= 0.25
Heads	+	Tails	=	0.5 × 0.5	= 0.25
Tails	+	Heads	=	0.5 × 0.5	= 0.25
Tails	+	Tails	=	0.5 × 0.5	= 0.25
				Total probabilities	= 1.00

However, if the coins are identical only three distinct events can be seen. The three distinct events are both heads, both tails and heads plus tails. The probabilities of the 3 types of event can be determined as shown below and once again these must add to unity.

This gives us a third rule:

> When events are not linked and are mutually exclusive their probabilities can be added to give the probability of the combined events occurring.

	p	
two 'heads'	0.25	
one 'heads' + one 'tails'	0.5	(0.25 + 0.25)
two 'tails'	0.25	
Total p = 1.00		

Similarly the probability of getting two consecutive 'heads' from a single coin is $0.5 \times 0.5 = 0.25$. This is because even though the two events are independent and do not influence each other, they are linked in that the first event must be successful before the second is considered.

Now consider the probabilities if **one** of the coins was weighted so that on average it only came up heads once in 20 throws or 5 per cent of all throws. 'Tails' would result in 19 out of 20 throws. Therefore

p'heads'(weighted) = 0.05
p'tails' (weighted) = 0.95
     Total p = 1.00

and the probabilities now become

Good coin		Weighted coin		probability	
Heads	+	Head	=	0.5×0.05	= 0.025
Heads	+	Tails	=	0.5×0.95	= 0.475
Tails	+	Heads	=	0.5×0.05	= 0.025
Tails	+	Tails	=	0.5×0.95	= 0.475
				Total	= 1.000

We can use these rules to investigate the probabilities that arise when batches are sampled in the search for defective components. If a process is under such good control that it produces no defective items and is thought to produce 5 per cent defective products when it has just started to go out of control, the probability of a single component, selected at random, being defective is 0.05. Hence the probability of it being good is 0.95, which is highly likely. So how do we know if the process is under control? For large batch sizes, removing a few samples does not significantly affect the proportions involved.

Let's take two samples from the same batch that contains 5 per cent defective items. The probability p of both being good is

$$p = 0.95 \times 0.95 = 0.90$$

For three samples the probability that they are all good is

$$p = 0.95 \times 0.95 \times 0.95 = 0.86$$

Notice how the probability of getting all good samples is going down the more we select. If this goes down the probability of the samples including at least one defective item must be going up. For two samples it is $(1 - 0.9) = 0.10$, for three samples it is $(1 - 0.86) = 0.14$, since probabilities of all linked events must sum to unity as usual.

If we take n samples the probability of them all being good is 0.95^n.

If we want to be 90 per cent certain that we have included at least one defective sample then we must reduce the probability of an **all good** sample to 0.10 by selecting N samples where

$$0.95^N = 0.1$$

This equation is best solved by taking logarithms of both sides to give

$$N \times \log_e 0.95 = \log_e 0.1$$

$$\therefore N = \frac{\log_e 0.1}{\log_e 0.95} = \frac{-2.30}{-0.05}$$

$$\therefore N \cong 45$$

to check $0.95^{45} = 0.1$

Hence for a batch thought to contain 5 per cent defective items then a sample of 45 items would include at least one defective item 90 times out of every 100 inspections made.

Notice that the absolute size of the batch is not part of the analysis. This is because it is just as easy to find 5 in 100 as 10 in 200 as 15 in 300, etc. In each case the defect level is 5 per cent. Hence the statistics of missing defective samples depends only on the numbers selected and the defect level, not the batch size. However, remember that the batch sizes have been assumed to be large enough such that the probabilities are not affected by the removal of the samples needed. In the above example where 45 samples are selected to ensure a 90 per cent chance of finding a defective item then the batch size would have to be at least 500.

Notice also that there is still a small chance that, even when 45 items are inspected, defective items are missed. Hence it may still be possible to wrongly assume that the process continues to function perfectly. This is the dilemma of inspection: finding the level of sampling that optimizes the inspection costs and the risk of letting batches that contain an unacceptable level of defects onto the market. This is a particularly difficult decision when the inspection is destructive and occurs at a late stage in production when the product's value is high.

Back at the supermarket, how many spot checks do we need to make in order to have a good chance of observing an intolerably long queue? Remember we can accept a defect level of two hours in fifty. That is 0.04. Therefore the probability of observing a shorter queue is 0.96 and we need to make sufficient spot checks (N) to ensure that this probability has fallen to a low level. Let's consider that this level is 10 per cent or 0.1. That means we will miss only one in ten of the unacceptable long queue events.

Therefore $0.96^N = 0.1$

and $N \ln 0.96 = \ln 0.1$

$N = \ln 0.1 / \ln 0.96$

$N = -2.30 / -0.04$

$N = 57.5$

Hence if spot checks are performed on a regular basis and only 1 in 60 reveals longer queue times than 5 minutes you can be reasonably sure that your check-out policy is being maintained.

Clearly the spot checks need to be made at sufficient intervals to allow queues to build up and not made at abnormal times such as just after opening. However if they are spaced out by very long intervals you run the risk of not detecting a deterioration in customer service soon enough.

Risk analysis in safety critical components and systems

It is not acceptable to utilize spot checks for components and services in the third of our categories. Here all products must be fully tested and no under-specification output allowed to reach the user. Such components and services are often linked to safety critical situations.

The same probability concepts can be used in a technique called **risk analysis.** Here there is no such thing as a risk-free component or service. All operations are assessed in terms of the probability of failure. For engineering components, they are all inspected to such a level that their chances of failure in their planned lifetimes are kept below a certain level.

For example, there are 1000 turbine blades in a particular gas turbine. The unexpected failure of any one of them can jeopardize the operation of the engine. If engines are designed to meet a probability of failure of once in 10 000 start-ups to full power, to what probability of failure, under identical conditions, must the blades be manufactured and inspected?

The performance of an individual blade is not linked to any other event. However, because each single blade can cause failure, their probabilities must be added together and equated to the overall probability of engine failure. In this way the probabilities can be thought of as being 'in series' since the integrity of the engine is linked to the integrity of each blade.

Hence $\quad \sum_{i=1}^{1000} p_i = \dfrac{1}{10\ 000}$

where i is the number of each blade and all blades have equal probability of failure p.

Therefore $\quad 1000p = \dfrac{1}{10\ 000}$

Hence $\quad p = \dfrac{1}{10\ 000\ 000} = 1 \times 10^{-7}$

This means that the blades themselves must be manufactured and inspected to this very demanding probability of unexpected failure. In this case it is a one in 10 million chance of unexpected failure during normal operating conditions. Such conditions are achievable with certain components and systems if it is possible to control manufacture and inspection to very high levels.

An alternative strategy is often used where such demanding standards cannot be achieved. The strategy is to duplicate or triplicate critical components within the system.

For example, diesel-powered electricity generators are used to provide occasional but essential power in the event of a power-cut. These are used in hospitals, department stores, factories and even high-rise residential buildings. However like all pieces of equipment they can fail to start up when needed. A probability can be placed on this event.

For one generator this might be

$p = 0.005$

This means that for 1 power cut in 200 you should expect the emergency generator to fail to start up, even though it has been regularly maintained.

This may be sufficient for a factory or department store but it may be considered too risky for a hospital. If a second generator is placed in parallel so that it attempts to start if the first one fails, then the probability of overall failure is now

$p = 0.005 \times 0.005 = 0.000025$

Now 25 in 1 000 000 or 1 in 40 000 power cuts is the chance of being without emergency electrical power because of an unexpected mains failure.

Even this may not be sufficient 'insurance' for the power to the cooling fluid pumps on a nuclear reactor or the hydraulic power for the control surfaces of an airliner. Here systems with similar probabilities may be triplicated to give a failure probability of

$p = 0.005^3 = 0.125 \times 10^{-6}$

This means failure of the system once in 8 million requests for service. This is still not zero but is now a very low risk of operational failure

104 Probability effects in projects

This type of analysis is a subject in its own right. It also requires engineers to examine hardware and software systems to check that replicated systems are fully independent of each other. The system components are also scrutinized by manufacturers to see if the intrinsic failure probabilities can be improved upon. Hence the trend in the aerospace industry is to replace mechanical and hydraulic control links with 'fly-by-wire' systems which appear to be fundamentally more reliable.

Risks in project implementation

The techniques of risk analysis are also combined with network analysis (Chapter 3) in many ways. For instance a critical path in a project might include a supplier providing goods or services on schedule. Experienced project engineers can often put a probability on a typical supplier failing to meet such crucial deadlines. If this seems too high then the cost of duplicating the supply route should be estimated and compared with the financial penalty within the whole project for delays in the critical path. This comparison may indicate that it is sensible to reduce risks by having alternative supply routes activated. More often, it means putting penalty clauses in contracts. However chasing up penalty payments is notoriously slow and unrewarding.

Political situations, workforce stability and even natural disasters should be taken into account when critical parts of projects are scrutinized for 'risk'. This is increasingly important as global approaches to manufacture become prevalent.

By means of example let us consider the risk in preparing meals. We all know that cookers break down, custard comes out lumpy, vegetables are not always as fresh as they appear on the outside, etc. Table 6.1 gives an estimate of the risks associated with the enjoyment of the two-course meal presented in the network diagram given again as Fig. 6.2. Calculate the risk for each of the 5 distinct paths through the diagram.

Notice again that there is no such thing as a risk-free event and that delay events also have an associated risk. The risks shown are for specific eventualities in this contrived example. In reality risk assessors deal with unspecified accumulated risks. For example, parking a car in an airport lot entails many risks from theft,

Probability effects in projects **105**

collision damage, etc. to even a minuscule risk of being hit by an aircraft. From an insurance point of view these risks would also be weighted in financial terms to cover the expected claims.

Table 6.1

	EVENT	DURATION	SPECIFIC RISK ASSESSMENT
A	DEFROST MEAT	20 MINS	meat rotten 1 in 1000 defrosts
B	PREPARE VEGS	14 MINS	vegs rotten 1 in 100 times
C	PREPARE PIE	20 MINS	pastry fails 1 in 200 times
D	BAKE PIE	45 MINS	pastry burns 1 in 300 times
E	MAKE CUSTARD	10 MINS	custard lumps 1 in 10 times
F	LAY TABLE	5 MINS	cutlery stolen 1 in 100 000 times
G	COOK VEG	15 MINS	pans boil dry 1 in 300 times
H	COOK MEAT	50 MINS	oven fails 1 in 6000 times
I	SERVE 1ST COURSE	3 MINS	server drops tray 1 in 12 000 times
J	EAT 1ST COURSE	20 MINS	boss telephones 1 in 500 times
K	EAT 2ND COURSE	10 MINS	unexpected guests 1 in 1000 times
L	DELAY PIE	28 MINS	dog eats fruit 1 in 2000 times
M	DELAY CUSTARD	83 MINS	cat drinks milk 1 in 500 times
N	DELAY TABLE LAY	65 MINS	kids loan out cutlery 1 in 100 000 times
O	DELAY COOK VEG	40 MINS	vegs spoil during delay 1 in 700 times

Table 6.2

	EVENT	RISK	REASON
A	DEFROST MEAT	1×10^{-3}	meat rotten 1 in 1000 defrosts
B	PREPARE VEGS	1×10^{-2}	vegs rotten 1 in 100 times
C	PREPARE PIE	5×10^{-3}	pastry fails 1 in 200 times
D	BAKE PIE	3.33×10^{-3}	pastry burns 1 in 300 times
E	MAKE CUSTARD	1×10^{-1}	custard lumps 1 in 10 times
F	LAY TABLE	1×10^{-5}	cutlery stolen 1 in 100 000 times
G	COOK VEG	3.33×10^{-3}	pans boil dry 1 in 300 times
H	COOK MEAT	1.67×10^{-4}	oven fails 1 in 6000 times
I	SERVE 1ST COURSE	3.33×10^{-4}	server drops tray 1 in 12 000 times
J	EAT 1ST COURSE	2×10^{-3}	boss telephones 1 in 500 times
K	EAT 2ND COURSE	1×10^{-3}	unexpected guests 1 in 1000 times
L	DELAY PIE	2×10^{-3}	dog eats fruit 1 in 2000 times
M	DELAY CUSTARD	2×10^{-3}	cat drinks milk 1 in 500 times
N	DELAY TABLE LAY	1×10^{-5}	kids loan out cutlery 1 in 100 000 times
O	DELAY COOK VEG	1.42×10^{-3}	vegs spoil during delay 1 in 700 times

106 Probability effects in projects

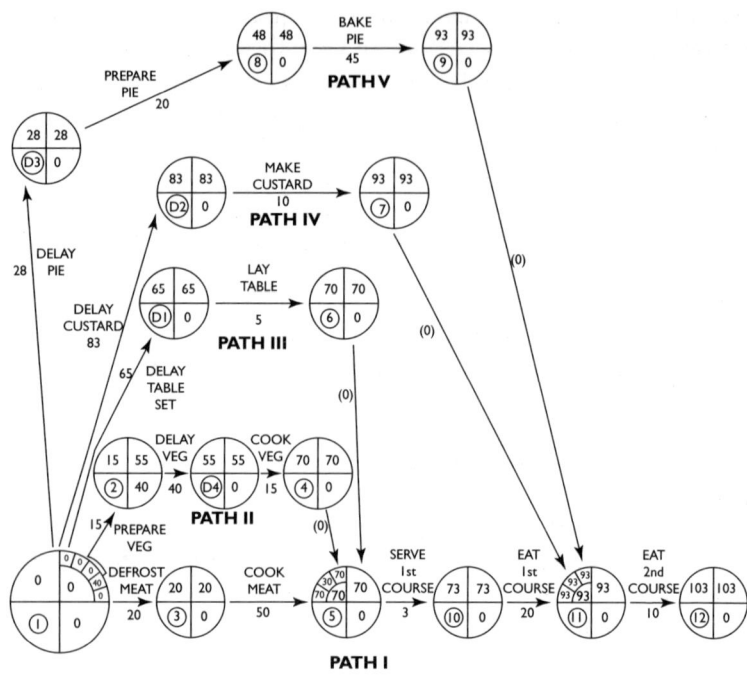

Figure 6.2 Network analysis of 2-course meal

For our example all we need to do is to examine the accumulated risk in each distinct path. The risks have been converted into decimal fractions as shown in Table 6.2.

The accumulated probabilities for the distinct paths can be calculated by adding together the individual probabilities for the specific events. The events are not linked but they all have the potential to ruin the meal. Hence the probability of each path not reaching a satisfactory end-point due to a failure of one event is the sum of the probabilities of each stage.

Path I is $1\times10^{-3} + 1.67\times10^{-4} + 3.33\times10^{-4} + 2\times10^{-3} + 1\times10^{-3} = 0.0045$
Path II is $1\times10^{-2} + 1.42\times10^{-3} + 3.33\times10^{-3} + 3.33\times10^{-4} + 2\times10^{-3} + 1\times10^{-3}$
$=0.01808$
Path III is $1\times10^{-5} + 1\times10^{-5} + 3.33\times10^{-4} + 2\times10^{-3} + 1\times10^{-3} = 0.0033$
Path IV is $2\times10^{-3} + 1\times10^{-1} + 1\times10^{-3} = 0.1030$
Path V is $2\times10^{-3} + 5\times10^{-3} + 3.33\times10^{-3} + 1\times10^{-3} = 0.0113$

Path IV is the most problematic. In addition to the problems of telephone calls and unexpected guests during the meal there is a relatively high probability of the lumpy custard ruining the meal. Hence 103 times in 1000 there will be a problem along this path. If lumpy custard does indeed ruin the whole meal this frequently, then you might consider duplicating the production event. If these duplicated batches of custard are independent of each other then the chances of them both being lumpy are reduced to $(1\times10^{-1}) \times (1\times10^{-1}) = 1\times10^{-2}$ and the probability of path IV falls from 0.1030 to 0.0130 and is comparable with other paths.

The overall probability of the meal being affected by any one single problem is not the addition of the probabilities of each path. The last three events are common to three paths and the last event is common to all paths as shown in Fig 6.2. Hence the overall probability is the summation of the probabilities of each event. This is equal to 0.1316 for single batch custard and 0.0416 for double batch custard. That is 13 in 100 and 4 in 100 respectively.

Following similar lines it is possible to predict if failure of particular events or pathways can affect project outcomes at an unacceptable frequency. Ideally all distinct paths through a project should have an equal chance of affecting the outcome and the overall probability of success should be acceptable.

When high risk events occur on non-critical paths it is not necessary to duplicate them if there is sufficient slack for them to be repeated.

Probability distributions

We have already referred to how probabilities of events may be linked to the frequency of particular events. In Chapter 5 the frequency of ambulance response times was linked to the probability of one arriving in a particular time interval. The histogram of hourly HGV traffic levels could have been converted to a probability distribution with the most probable of the hourly events being HGVs at four vehicles per hour.

The concept of a distribution of probabilities for the same event is particularly important. It introduces the idea of a spread of outcomes. Some of the outcomes will be more frequently observed than others and all feasible events are possible. For instance, as shown in Fig. 5.15, it is most likely that the ambulance you called will arrive in 20–25 minutes. However it is possible that it might arrive in under 5 minutes.

Hence individual response times are not a sensible comparison of two ambulance stations. It would be necessary to examine the spread of response times to make a more meaningful comparison. Before we attempt such comparisons let's introduce some basics. There are several types of probability distribution.

1. Rectangular distributions

Here the probability is uniform throughout the range. The simplest example of this would be the probability of the scores 1→6 when a single dice is thrown. This is shown in the histogram in Fig. 6.3. The scores of 1, 2, 3, 4, 5 and 6 are all feasible outcomes of this event and if a properly balanced dice is correctly thrown all these outcomes have the same probability of occurring. That is 1 in 6 = 0.1666.

Once again the probability is represented by the area and the total area is 1.

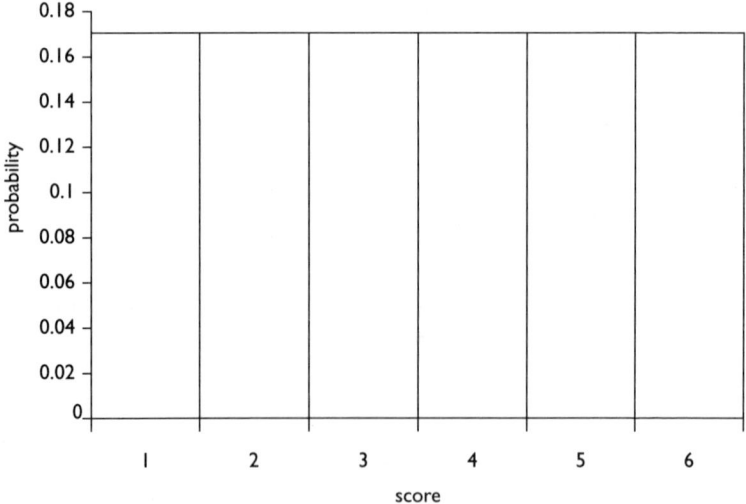

Fig. 6.3 The uniform spread of probability in a rectangular distribution

2. Symmetrical binomial distributions

These result wherever distributions involve 50/50 individual events. The simplest of such events is the distribution of events produced by the sequential tossing of a coin.

The first toss has two possible outcomes and we can cover these by monitoring the number of 'heads' produced.

1 trial No. of 'heads' 0,   1
             probability   1/2, 1/2

This represents a rectangular distribution but if we now go to two trials we get:

2 trials No. of 'heads' 0   1,   2
              probabilities   1/4   1/2 1/4

Notice now that the outcome of '1 heads' is now more probable. This is because there are now four possible outcomes T + T, H + T, T + H, H + H, and two of these events are indistinguishable since they each produce one 'heads'. Hence the event of one 'heads' in two tosses is twice as frequent and has double the probability.
With 3 trials the distribution becomes:

3 trials No. of 'heads' 0,   1,   2,   3
              probabilities   1/8, 3/8, 3/8, 1/8

If p is the probability of a 'heads' and q is the probability of a 'tails', three consecutive 'heads' has the probability $p^3 = 1/2^3 = 1/8$. Likewise no 'heads' has $q^3 = 1/2^3 = 1/8$. In between these extremes the odds of getting a situation with one 'heads' is $p \times q \times q = 1/8$ and two 'heads' is $p \times p \times q = 1/8$.

However, there are three situations in which the single 'heads' can appear and three ways for the 'two heads' to appear so their overall probabilities are both $3 \times 1/8 = 3/8$.

By similar reasoning four trials gives:

4 trials No. of 'heads' 0,   1,   2,   3,   4
              probabilities   1/16, 4/16, 6/16,   4/16, 1/16
5 trials No. of 'heads' 0,   1,   2,   3,   4,   5
              probabilities   1/32, 5/32, 10/32, 10/32, 5/32, 1/32

Notice how the probability distributions are getting distorted to favour the central values. However they remain perfectly symmetrical due to the equal probability of the contributing event, i.e. $p = q$ and $p + q = 1$.

3. General binomial distributions

In order to generalize, the probabilities of the individual events remain as $p + q = 1$ but p does not need to be equal to q.

The distribution is created by considering there to be n possible outcomes and the probability of any particular outcome r, in the range 0 to n, is given by:

$$\frac{n!}{(n-r)!r!} q^{n-r} p^r$$

If this symbolism is not clear it is explained in the next example.

For $r = 0$ this expression reduces q^n, since 0! is defined to equal 1.

For r = n this expression reduces p^n, since $q^0 = 1$.

Hence the extreme values can be calculated as before. In between, the probability depends on the relative sizes of p and q and the position r in the sequence of all possible outcomes from 0 to n.

Probability effects in response rate calculations

Many projects require data to be collected or trials to be completed. Response levels to mailshots, questionnaires and market surveys are never 100%. Hence it is a good idea to be able to estimate how many extra samples are needed in order to achieve the required amount of data. Similarly it is often necessary to overbook services in order to ensure that hotels, planes, theatres, etc. are filled to economic levels. However it is best not to over-do-it because it will then be probable that you have some very disappointed customers. Sales targets should also include an estimate of the probability that initial enquiries do not result in clinched deals.

For example if we were interested in assessing public reaction to a new software product but did not want to provide the hardware, we would need to know the level of access of potential testers to a suitable computer. If this could be established we could then work out the probabilities of particular access levels from typical test groups. We will assume that this access is the only barrier to getting the product assessed and that 1 in 5 people has access to a suitable computer. Hence $p = (1/5)$ and $q = (4/5)$.

If a typical group size(n) was 20, then the probability of getting 20 software assessments within one group would be p^n.

Hence $\quad P_{20} = p^n = \left(\frac{1}{5}\right)^{20} = 0.2^{20}$

therefore $P_{20} = 1 \times 10^{-14}$ i.e. exceedingly small

The individual probability that an individual does not have access is $1 - 1/5 = 0.8 = q$.
Hence the probability of zero access from a group of 20(Q_0) is given by q^n

$Q_0 = (0.8)^{20}$

therefore $Q_0 = 0.0115$ just over 1 in 100 times

These are the two extreme values. In between these the calculation of the probabilities is slightly more complex. However these are the most useful results and so let's look at how these values may be estimated.

For example, we may be interested in the probability of getting 25 per cent of the group with access. Since the group size is 20 that means 4 people with access. One way of achieving this would be if the first 4 had access and the remaining 16 did not. The probability of this single occurrence may be calculated as before.

The probability of the first four having access is given by

$P_{4y} = (0.2)^4$

This probability is reduced because the next 16 must be those without access. Their individual probability of no access is 0.8 as before and as each one is chosen the overall probability goes down accordingly. The overall probability for this particular situation is given by
$P_{4y16n} = (0.2)^4 \times (0.8)^{16}$

However, there is nothing special about this particular way of finding 4 people with access. It was just as convenient to do the calculations for the situation where the first 4 people have access. There are many ways of finding 4 people in a group of 20 and each one has equal probability. Hence

Total probability = Individual probability × Number of equivalent situations.

Now a general formula for the number of distinct ways of selecting, arranging, finding or dispersing r items amongst a total of n items is written as $_nC_r$ and is given by

$$_nC_r = \frac{n!}{r!(n-r)!}$$

where the symbol $n!$ means 'n factorial' and is simply $1\times2\times3\times4.......\times n$ where n is a positive whole number. There are $n!$ permutations when n items are arranged amongst themselves and none of these are distinct. However, when items are mixed in, many of the arrangements are distinct but some of these arrangements are simply created by shuffling the selected items (r) amongst themselves and the ($n - r$) unselected items amongst themselves. Hence the number of distinct arrangements is determined by dividing the total by $r!$ and ($n - r$)! respectively.

Hence the number of distinct ways of selecting 4 from 20 is

$$_{20}C_4 = \frac{20!}{4! \times (20-4)!}$$

$$= \frac{20!}{4! \times 16!}$$

$$= \frac{20\times19\times18\times17\times16\times15\times14\times13\times12\times11\times9\times8\times7\times6\times5\times4\times3\times2\times1}{(4\times3\times2\times1)\times(16\times15\times14\times13\times12\times11\times10\times9\times8\times7\times6\times5\times3\times2\times1)}$$

$$= \frac{20\times19\times18\times17}{4\times3\times2\times1}$$

$$= 4845$$

That means there are still over four thousand eight hundred distinct ways of arranging 4 things amongst 20 locations, not counting the ones where 'r items' and others are shuffled amongst themselves.

Hence the total probability of the situation where 4 people amongst 20 have access is

$$p_{4t} = 4845 \times 0.2^4 \times 0.8^{16}$$

$$\therefore p_{4t} = 0.2182$$

Similar calculations can be carried out for other levels of access and the results are shown in Table 6.3. Also shown is the accumulative probability for the level of access plus all levels below. Notice how this rises quickly and then tails off as the accumulated probability approaches 1. This means that it would be very rare to find groups of 20 with more than 13 in the group with suitable access. This data can also be plotted on a frequency distribution diagram as shown in Fig 6.4.

Table 6.3

r	Probability of r out of 20	Accumulated probability for less than r out of 20
0	0.0115	0.0115
1	0.0576	0.0692
2	0.1369	0.2061
3	0.2054	0.4114
4	0.2182	0.6296
5	0.1746	0.8042
6	0.1091	0.9133
7	0.0545	0.9679
8	0.0222	0.9900
9	0.0074	0.9974
10	0.0020	0.9994
11	0.0005	0.9999
12	0.0001	1 approx.
13	<0.0001	1 approx.
14	<0.0001	1 approx.
15	<0.0001	1 approx.
16	<0.0001	1 approx.
17	<0.0001	1 approx.
18	<0.0001	1 approx.
19	<0.0001	1 approx.
20	<0.0001	1 approx.

Probability effects in sales teams planning

This type of data could be generated and used in sales situations. For example if couples are approached and talked to in batches of 20 about 'time-share' or private pension schemes or double-

114 Probability effects in projects

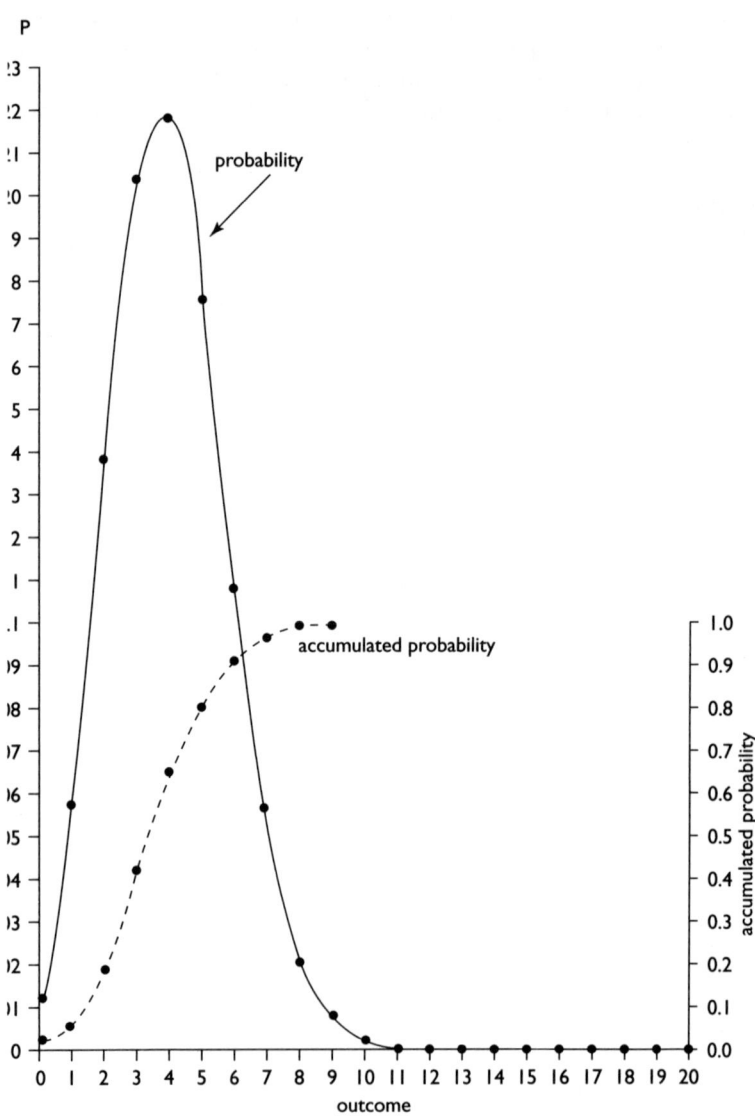

Fig. 6.4 Individual and accumulated probabilities as shown in Table 6.3

Probability effects in projects

glazing, etc., then if you know the typical rate at which they are interested in taking things to the next stage, you can pitch your support staff levels accordingly. If the individual rate of interest is 1 in 5 the data is identical to that given in Fig. 6.4.

Hence the dotted curve would indicate that 60 per cent of all possibilities would be immediately covered by having resources to handle just 4 couples in the next stage, 90 per cent of all possibilities could be handled by staff to deal with 6 couples and 99 per cent cover would be obtained by staff to handle 9 couples. There may be the odd occasion when more couples want to proceed to the next stage.

This type of distribution can be described as skewed since the highest probability is not associated with mid-range values on the horizontal axis. This also means that the accumulated probability also reaches 50% or 0.5 at an event outcome much lower than the mid-range value.

Probability effects in mass production

You will recall that the fear of missing a single defective sample in a batch suspected of containing 5 per cent defective items caused the sample size to be increased to 45 to give a probability of only 0.1 that they were all good samples.

A common situation is to relax specifications and tolerate processes that produce some defectives. This may again be 5 per cent. The general binomial expression is for the probability (P_r) of a certain outcome r from n possible outcomes where the contributing probabilities are p and q is:

$$P_r = \frac{n!}{(n-r)!r!} q^{n-r} p^r$$

It can be used to estimate probability effects in samples taken from production batches containing specific defect levels which are acceptable. For instance you might heat up drying kilns so quickly to season wood that the product contains 5% cracks, or cool down pottery furnaces quickly and tolerate the 5% cracked crockery produced. With $p = 0.05$, $q = 0.95$ and a sample size of 45 we can estimate these probabilities of finding a particular level of defective items in the sample.

For 0 defects $r = 0$

$$P_0 = \frac{45!}{45!\,0!} \; q^{45} \times p^0$$

now $x^0 = 1$ and $0!$ is defined as being equal to 1 as before

$\therefore P_0 = q^{45} = 0.95^{45} = 0.1$

for 1 defect $r = 1$

$$P_1 = \frac{45!}{44! \times 1!} \; 0.95^{44} \times 0.05^1$$

$= 0.236$

for 2 defects $r = 2$

$$P_2 = \frac{45!}{43! \times 2!} \; 0.95^{43} \times 0.05^2$$

$$= \frac{45 \times 44}{2} \times 0.95^{43} \times 0.05^2$$

$= 0.272$

for 3 defects

$$P_3 = \frac{45!}{42! \times 3!} \; 0.95^{42} \times 0.05^3$$

$$= \frac{45 \times 44 \times 43}{6} \times 0.95^{42} \times 0.05^3$$

$= 0.206$

or 4 defects

$$P_4 = \frac{45!}{41! \times 4!} \; 0.95^{41} \times 0.05^4$$

$$= \frac{45 \times 44 \times 43 \times 42}{24} \times 0.95^{41} \times 0.05^4$$

$= 0.1136$

for 5 defects

$$P_5 = \frac{45!}{40! \times 5!} \, 0.95^{40} \times 0.05^5$$

$$= \frac{45 \times 44 \times 43 \times 42 \times 41}{120} \times 0.95^{40} \times 0.05^5$$

$$= 0.049$$

Now with a sample of 45 from a process that produced 5 per cent defects, on average, you would expect about two defects. However, as you can see from the above calculation there are significant probabilities of getting anywhere in the range 0–5 defects.

This poses a major dilemma of process control in mass production.

When do you stop a mass production line?
If you got five defects, is it due to the normal spread of data (remember $P = 0.049$ for 5) or is it symptomatic that the process is going out of control and the next few batches are going to contain an unacceptable defect level with consequential commercial problems?

It can be a 'no-win' situation. Hence the area of statistical process control where such problems are considered in great depth is an emerging technology and is considered in the next chapter.

Problems

1. A closed bag contains 10 white balls and 10 black balls. What are the chances of drawing three consecutive black balls if the ball is replaced after each draw and the bag shaken up?

2. A bag contains 10 white balls and 10 black balls. If three balls are drawn in succession without looking at them what is the overall probability that they will all be black?

3. Russian 'lager' roulette is played by shaking up a sealed can of beer and placing it with five unshaken identical cans. The cans are then gently shuffled. The player chooses one and opens it under his/her chin. What are the chances of not being sprayed with beer if you play the game?

(1) once?

(2) twice?

(3) three times?

(4) six times?

using a fresh six-pack on each occasion.

4. Two cards are drawn at random from a pack of 52 card deck. What is the probability that they are the King of Clubs and the Jack of Diamonds?

5. My sock drawer contains 10 white socks, 10 black socks and 10 green socks randomly arranged. If I select them in the dark what is the probability of getting a matched pair if

   (a) two are selected?

   (b) three are selected?

   (c) four are selected?

6 The UK National Lottery has a jackpot when 6 different numbers are selected in the range 1 to 49. Find the odds of winning.

7. If the proportion of cars fitted with driver's air bags is 20 per cent what is the probability that **more than half** the cars in a pile-up of six are fitted with driver's air bag protection.

Answers

1. On each occasion the probability of drawing a black ball is 10/20 or 1/2. The three draws are linked in overall probability terms but each draw does not influence the next one.
   Hence the overall probability $= 1/2 \times 1/2 \times 1/2$
   $= 0.5 \times 0.5 \times 0.5$
   $= 0.125$ or 1 in 8

2. The logic is the same as in question 1, however, if the balls are not replaced the probability of choosing consecutive blacks goes down. The probability of the first ball being black is $10/20 = 0.500$. However if we have withdrawn a black ball the probability of getting a second black ball is now $9/19 = 0.473$. The prob-

ability for a third black ball is 8/18 = 0.444.
Now the probability of three consecutive black balls is

$$= \frac{10}{20} \times \frac{9}{19} \times \frac{8}{18}$$

$$= \frac{8}{76} = 0.105$$

(Check 0.5 × 0.473 × 0.444 = 0.105)

3. The probability of choosing the 'live' can each time is 1/6 so the odds of choosing an unshaken can are 5/6. Hence the odds of keeping a clean face are:

one round = 5/6 = 0.833

two rounds = 5/6 × 5/6 = 0.694

three rounds = 5/6 × 5/5 × 5/6 = 0.579

six rounds = $(5/6)^6$ = 0.335

4. The first card drawn may be either one of the two named cards. Hence the probability is 2/52. The next card **must** be the other one and the deck has now decreased to 51 cards hence the probability is 1/51 so the overall probability is

$$\frac{2}{52} \times \frac{1}{51} = \frac{2}{2652} = \frac{1}{1326}$$

$$= 0.000754$$

5. (a) For a selection of just two socks let's consider the probability of getting a pair of white socks. The probability that the first sock is white is 10/30. The probability that the second sock is also white is 9/29 since a white sock has already gone and the total is also reduced by one. Hence the overall probability for this linked event is

$$\frac{10}{30} \times \frac{9}{29} = \frac{3}{29} = 0.103$$

However, the question only asked about a pair and this could occur for green and black socks. Those possibilities are additional to that for the white socks. Hence the overall probability

$$= 3 \times \frac{3}{29} = \frac{9}{29} = 0.309$$

Another way to look at it is to say you are certain about the first sock whatever the colour. Hence its probability is 1. Now to get another of the same colour you must pick one of the remaining 9 from the 29 in the drawer. Hence the overall probability is 1 × 9/29 as before.

c) When three socks are chosen the situation is more involved. For a pair of white socks there are four ways of achieving a pair as shown below for white socks (W) or non-white (N), along with the probability of the particular combination.

Socks	Probabilities	
W + W + N	$\frac{10}{30} \times \frac{9}{29} \times \frac{20}{28}$	= 0.07389
W + N + W	$\frac{10}{30} \times \frac{20}{29} \times \frac{9}{28}$	= 0.07389
N + W + W	$\frac{20}{30} \times \frac{10}{29} \times \frac{9}{28}$	= 0.07389
W + W + W	$\frac{10}{30} \times \frac{9}{29} \times \frac{8}{28}$	= 0.02955
Sub total probability for a pair of white socks		= 0.25122

Now this situation is triplicated because we can also get a pair by having black or green socks.

Hence the overall probability of getting a pair of **any** colour when three are chosen is 0.25122 × 3 = 0.75366.

(c) When four are chosen we can be certain of getting a pair because a pair is created if three of a kind or a pair plus an odd sock are chosen. The only other situation is when three different colours are chosen. The fourth one **must** then create a pair with one of them. Hence the probability is 1.

6. The odds of winning are calculated by finding the number of ways of selecting **distinct** arrangements of the six numbers from the 49 numbers since the order in which the balls are selected is unimportant. This is written as $_{49}C_6$

$$= \frac{49!}{(49-6)!6!} = \frac{49 \times 48 \times 47 \times 46 \times 45 \times 44}{6 \times 5 \times 4 \times 3 \times 2 \times 1}$$

$$= 13\,983\,816$$

Hence the odds of winning are approximately 14 million to one against for a single entry of six different numbers.

7. The probability of a fitted air bag = 0.2

   Therefore the probability of no airbag = 0.8

   The probability required is

$$P_{4 \text{ from } 6} + P_{5 \text{ from } 6} + P_{6 \text{ from } 6}$$

$$P_{4 \text{ f } 6} = 0.2^4 \times 0.8^2 \times {}_6C_4$$

$$= 0.2^4 \times 0.8^2 \times \frac{6!}{2!4!} = 0.01536$$

$$P_{5 \text{ f } 6} = 0.2^5 \times 0.8^1 \times {}_6C_5$$

$$= 0.2^5 \times 0.8 \times \frac{6!}{1!5!} = 0.001536$$

$$P_{6 \text{ f } 6} = 0.2^6 = 0.00064$$

Total P = 0.0175

Hence the probability of more than half the cars having air bags = 0.017 which is less than 2 per cent!

Objectives

After studying this chapter you should be able to do the following.

1. Calculate combined probabilities for linked events and unlinked events.
2. Distinguish between rectangular and binomial probability distributions.
3. Give the formula for a general binomial distribution and use it to calculate individual and accumulated probabilities of event outcomes.

4. Explain how probability theory can be used in risk assessments and in conjunction with Network Analysis.
5. Calculate the probabilities of particular outcomes in response to rate assessments, surveys, samples and similar situations that produce binomially distributed event outcomes.

7 Standard deviations and normal distributions

Aims

- To introduce the standard deviation as a means of assessing the spread of values about a mean
- To introduce the concepts of means and standard deviations of probability distributions
- To introduce the 'normal distribution' concept
- To use the normal distribution as an approximation to measured distributions in project data
- To use standard deviations and standard errors for project and process control purposes

Standard deviation

The last chapter highlighted the uncertainties when data is distributed in a binomial fashion. This means that different outcomes have different probabilities.

Such distributions are very common. For example, in electronic component manufacture, many simple components are manufactured to have particular target values. Resistors and capacitors are made to have particular resistances and capacitances respectively. However variations in the production processes mean that their actual values are distributed about a mean. Such distributions occur for discrete data, grouped data and continuous data.

Unfortunately it is possible to have two distributions with identical ranges, arithmetic means, medians and mid-range values. However the same data can be dispersed about the mean in different ways as shown in Fig. 7.1.

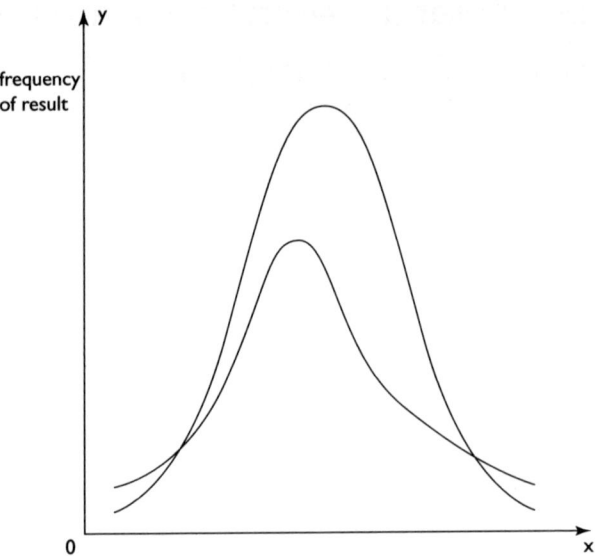

Figure 7.1 Two distributions with equal ranges, mid-values, and means

In Fig. 7.2 two other distributions also have the same mode! However if these distributions represented output from different electronics factories or processes then the one with the least dispersion about the mean would be favoured. (other things being equal). This would be because less 'out-of-range' components would have to be rejected. For situations like this the **standard deviation** of the data is a most useful measure. It is a little tricky to develop a way to find the important feature of the spread of values. If we add up all the deviations of individual data points from the arithmetic mean, we already know that they cancel out since

$$\Sigma(x - \bar{x}_a) = 0$$

Hence this cannot be used as a measure of the dispersion. An alternative is to square all the values and take the arithmetic mean i.e.

$$\frac{\Sigma(x - \bar{x}_a)^2}{n}$$

This would give a measure of the deviation and it is called the variance of the data.

The **standard deviation**[*] can be found by taking the positive square root of this variance term.

Hence
$$S = +\sqrt{\frac{\Sigma(x_i - \bar{x}_a)^2}{n}}$$

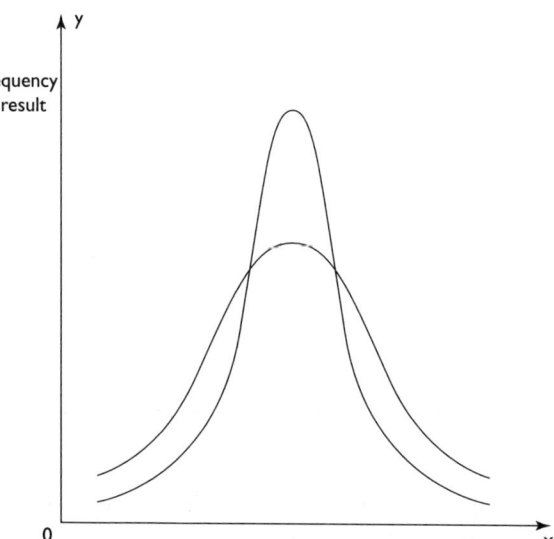

Figure 7.2 Two distributions with equal ranges mid-values, means and modes.

When grouped data or data from frequency distributions is used the formula becomes

$$S = +\sqrt{\frac{\Sigma f_i(x_i - \bar{x}_a)}{n}}$$

---

[*]In more advanced study of statistics a modification of the formula for standard deviations is used where

$$S = +\sqrt{\frac{\Sigma(x_i - \bar{x}_a)^2}{n-1}}$$

However, this is not necessary in this text.

where f_i is the frequency of data in the group and $n = \Sigma f_i$

One of the most common uses of standard deviations is in examinations and assessments. Shown in Table 7.1 are the marks out of 10 of a group of students taking a mid-term test. Notice that the marks have been grouped. Now the average mark can be stated as 4 with a standard deviation of 1.414.

Table 7.1

Marks x_i	No. of students f	fx_i	$x_i - \bar{x}_a$	$(x_i - \bar{x}_a)^2$	$f(x_i - \bar{x}_a)^2$
0	1	0	−4	16	16
1	8	8	−3	9	72
2	28	56	−2	4	112
3	56	168	−1	1	56
$\bar{x}_a = 4$	70	280	0	0	0
5	56	280	1	1	56
6	28	168	2	4	112
7	8	56	3	9	72
8	1	8	4	16	16
	256	1024			512

$$n = \Sigma f = 256, \quad \bar{x}_a = \frac{\Sigma fx_i}{n} = \frac{1024}{256} = 4$$

$$S = \sqrt{\frac{\Sigma f(x - \bar{x}_a)^2}{n}} = \sqrt{\frac{512}{256}} = \sqrt{2} = 1.414...$$

Standard deviations and normal distributions

This indicates an educationally sound spread of results. As you can see from the data the modal value is 4/10 and there is a very symmetrical distribution about 4.

Educationalists like to see this type of spread. If everyone got between 4 and 5 marks the standard deviation would be very much smaller and the test criticized for not discriminating between performances. A very even distribution that tended towards a rectangular distribution with the same number of students in each 'marks' category would also be criticised for being an abnormal response. Here the standard deviation would be extremely high.

In project work the main use of the standard deviation then is to compare the distribution of data about a mean value. If you were to assess the average disposable incomes of different groups of people the value of the standard deviation would distinguish how that income was distributed. A small standard deviation would indicate a large proportion close to the average value. A larger value would indicate a more even distribution. This would probably mean more fairly rich people but also more fairly poor people. Hence your marketing strategy would be adapted accordingly.

Means and standard deviations of probability distributions

It is quite straightforward to work out the arithmetic mean of a **frequency distribution**. If the data is simply listed with all the individual results you would simply use the formula as usual.

$$\bar{x}_a = \frac{\sum_{i=0}^{i=n} x_i}{n}$$

This can be applied to continuous or grouped data. For grouped data an alternative way would be to use the frequency number 'f' in the equation where x is the average value of the group. This was done in the earlier calculation of standard deviation of class marks using this expression.

$$\frac{\Sigma fx}{n}$$

This simply multiplies the grouped data levels by the frequency of their occurrences.

However n is simply a constant and so this can be written as.

$$\Sigma \frac{f}{n} x$$

Now $\frac{f}{n}$ is simply the probability of a particular value occurring (P).

Therefore the mean value (μ) of data expressed as a **probability distribution** is given by

$$\mu = \Sigma Px$$

It is slightly more abstract to associate a physical meaning to this mean of a probability distribution. If the distribution is number of heads found in successive tosses of a coin, as discussed earlier, it could be regarded as the **average** number of heads per trial.

This is because the number of successes would be distributed as usual. Now any distribution about a mean has a standard deviation. Following similar reasoning the **standard deviation of a probability distribution** is

$$\sigma = \sqrt{\left[\Sigma P(x-\mu)^2\right]}$$

where σ represents the standard deviation of a probability distribution with a mean of μ and P is the probability of an individual outcome x within the distribution.

Alternative formulae for the mean and standard deviation of a probability distribution

If we have an event where the probabilities are such that $p + q = 1$ where p is the probability of a successful outcome then this produces the following distributions (which may be calculated using the general formulae) when different numbers of outcomes ($x = 0$ to n) are possible

(i)

x	0	1
P	q	p

(ii)

x	0	1	2
P	q^2	$2qp$	p^2

(iii)

x	0	1	2	3
P	q^3	$3q^2p$	$3qp^2$	p^3

For the mean of each distribution μ
using the formula $\mu = \sum Px$

in situation (i) $\mu = \sum Px = (0 \times q) + (1 \times p)$, $\therefore \mu = p$, when $n = 1$

in situation (ii) $\mu = \sum Px = (0 \times q^2) + 2qp + 2p^2$
$$= 2p(q+p)$$
but $q+p=1$

$\therefore \mu = \sum Px = 2p,$

$\therefore \mu = 2p$, when $n=2$.

in situation (iii)

$\mu = \sum Px = (0 \times q^3) + 3q^2p + 6qp^2 + 3p^3$
$= 3p(q^2 + 2qp + p^2)$
but $q^2 + 2qp + p^2 = (q+p)^2$
$\therefore q^2 + 2qp + p^2 = 1^2 = 1$
$\therefore \mu = \sum Px = 3p,$
$\therefore \mu = 3p$, when $n = 3$

Can you see the trend? Well it continues and so we can write the general result that

$\mu = np$.

This implies that the mean of a probability distribution is the intrinsic probability of success times the number of possible outcomes.

Stated in words this relationship is even more obvious. If we keep selecting 300 adults at random the number of men in the group will not be the same each time. However the mean of the distribution of numbers of men will be determined by the probability of finding men in the sample and the size of the sample. Now clearly $p = 1/2$ and $n = 300$ (group size).

So $\mu = 1/2 \times 300 = 150$

Hence the average number of men in the sample is the mean of the distribution which in this case is 150.

So in summary if you take groups of 300 adults at random, **on average** you will select 150 men. However the actual values of the outcomes will be distributed about this mean.

For the standard deviations of each of the above distributions, σ

using the formula $\sigma = \sqrt{\Sigma P(x - \mu)^2}$

in situation (i) $\mu = np$, ∴ $\mu = p$

$\sigma^2 = \Sigma P(x-\mu)^2 = qp^2 + p(1-p)^2$

but $1 - p = q$

∴ $\sigma^2 = qp^2 + pq^2$

$= qp(p + q)$

$= qp$ (since $p + q = 1$)

∴ $\sigma = \sqrt{qp}$ when $n = 1$

in situation (ii) using $\mu = np = 2p$ and $q + p = 1$ therefore $1 - p = q$

$\Sigma P(x - \mu)^2 = q^2(0 - 2p)^2 + 2qp(1 - 2p)^2 + p^2(2 - 2p)^2$

$= 4p^2q^2 + 2qp(1 - p - p)^2 + p^2(1 - p + 1 - p)^2$

$= 4p^2q^2 + 2qp(q - p)^2 + p^2(2q)^2$

$= 8p^2q^2 + 2qp(q - p)^2$

$= 2qp(4qp + q^2 - 2qp + p^2)$

$= 2qp(q^2 + 2qp + p^2)$

$= 2qp(q + p)^2$

$= 2qp$ (since $p + q = 1$)

∴ $\sigma = \sqrt{\Sigma P(x - \mu)^2} = \sqrt{2qp}$ when $n = 2$.

in situation (iii) using $\mu = 3p$

$\sum P(x - \mu)^2 = q^3(0 - 3p)^2 + 3q^2p(1 - 3p)^2 + 3qp^2(2 - 3p)^2 + p^3(3 - 3p)^2$
$= 9p^2q^3 + 3q^2p(1-p-p-p)^2 + 3qp^2(1-p+1-p-p)^2 + p^3(1-p+1-p+1-p)^2$
remembering $1 - p = q$
$= 9p^2q^3 + 3q^2p(q - 2p)^2 + 3qp^2(2q - p)^2 + p^3(3q)^2$
$= 3pq(3pq^2 + q(q - 2p)^2 + p(2q - p)^2 + 3p^2q)$
$= 3pq(3pq^2 + q(q^2 - 4qp + 4p^2) + p(4q^2 - 4qp + p^2) + 3p^2q)$
$= 3pq(3pq^2 + q^3 - 4q^2p + 4p^2q + 4q^2p - 4qp^2 + p^3 + 3p^2q)$
$= 3pq(q^3 + 3q^2p + 3p^2q + p^3)$
$= 3pq(q + p)^3$
$= 3pq$ (since $q + p = 1$)
$\therefore \sum P(x - \mu)^2 = 3pq$
$\therefore \sigma = \sqrt{\sum P(x - \mu)^2} = \sqrt{3pq}$ when $n = 3$

Can you see a similar trend this time? Once again it continues as n increases and we can write the result that

$\sigma = \sqrt{npq}$

Summarizing this means that for a probability distribution where p and q are the individual probabilities that contribute to the outcome and n is the number of possible outcomes the mean $\mu = np$ and the standard deviation $\sigma = \sqrt{npq}$.

From the symmetric binomial distribution to the 'normal' probability distribution

You might have noticed that the probabilities of the particular outcomes in binomial distributions are simply the terms involved in the expansion of $(p+q)^n$. This gets cumbersome for large n values.

Just as we 'scaled down' the arithmetic means earlier in Chapter 5, the values plotted on the horizontal axis of our probability distributions can be scaled down by plotting them as X values where

$$X = \frac{x - \mu}{\sigma}$$

where μ is the mean of the probability distribution and σ is the standard deviation.

If we had a perfectly symmetrical distribution resulting from the expansion of

$$(1/2 + 1/2)^n$$

then this procedure would conveniently centre the distribution about the zero on the horizontal axis. However, it would also reduce the area under the histogram by a factor of $1/\sigma$. Hence the values on the vertical axis, that is the probability P, should be increased accordingly to a value Y where

$$Y = P\sigma.$$

Now the area under the histogram remains equal to 1 like all the probability distributions considered. The scaling of the data becomes important as n gets bigger and bigger. As n increases, it would also be possible to divide the histogram into smaller and smaller sections as shown in Fig. 7.3. If the centres of the bar tops are eventually replaced by a continuous curve Fig. 7.4 results. This is known as the normal probability distribution and many measurements are distributed around their mean according to this curve. It has the equation

$$y = \frac{1}{\sqrt{2\pi}} e^{-\frac{1}{2}x^2}$$

It is important to recognize that this now represents a continuous function. The y values can take **any** level, not just the discrete levels produced by a theoretical distribution or the levels recorded in grouped data distributions. Hence the equations for the mean and standard deviation as given earlier are not valid here. However the area under the curve is still equal to one.

The maximum value on the y axis occurs when $x = 0$ and is

$$y = \frac{1}{\sqrt{2\pi}} e^0 = \frac{1}{\sqrt{2\pi}} \quad \therefore y = 0.4$$

Standard deviations and normal distributions

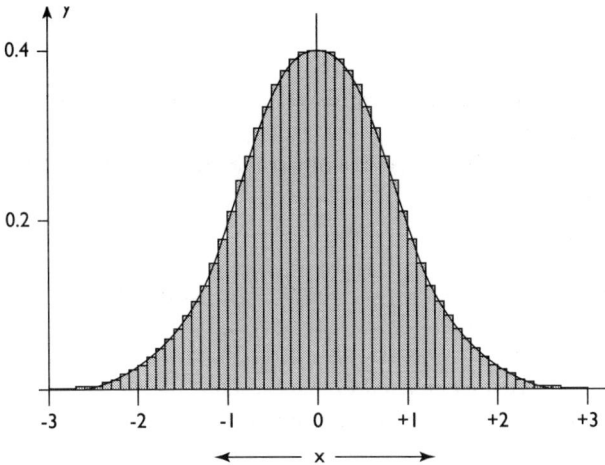

Figure 7.3 Probability distribution of $(1/2 + 1/2)^n$ centred on zero by scaling

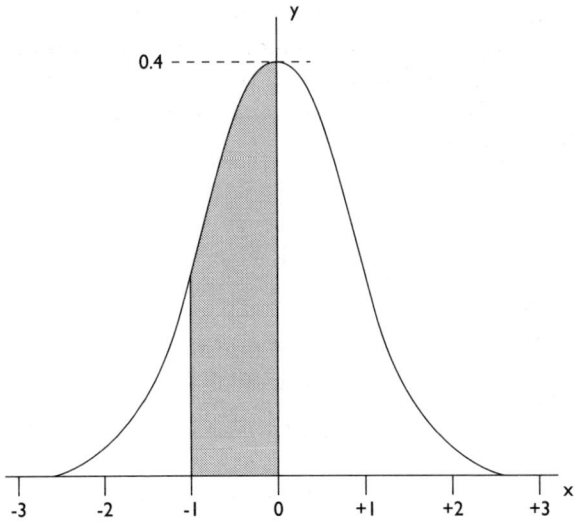

Figure 7.4 Continuous curve replacing histogram of Figure 7.3

As you will see almost all the values occur within a range ±3, on the horizontal axis (Fig. 7.4).

Remembering that the horizontal axis represents $\dfrac{x-\mu}{\sigma}$

$\dfrac{x_{max}-\mu}{\sigma} = 3$ is close to the upper limit of the data

$\therefore x_{max} \approx \mu + 3\sigma$

and $\dfrac{x_{min}-\mu}{\sigma} = -3$ is close to the lower limit of the data

$\therefore x_{min} \approx \mu - 3\sigma$

Hence there is very high probability that the data falls within ±3 standard deviations of the mean value in a normal distribution

The other key concept is that **the area under the curve between two values on the x axis represents the probability that any value of the data plotted falls between these values.**

Hence the shaded area on Fig. 7.4 represents the probability that any value in a normal distribution falls between

$\dfrac{x-\mu}{\sigma} = 0$, i.e. $x = \mu$

and $\dfrac{x-\mu}{\sigma} = -1$, i.e. $x = \mu - 1\sigma$

Therefore the area represents the probability of a value falling within the range of one standard deviation below the mean.

The areas involved under such curves can be found by the mathematical process of integration and Table 7.2 shows the symbolism and the results. Because the curve is symmetrical only the results from zero to 3.99 on the horizontal axis are shown. As you can see from the first entry, the area under the curve is exactly 0.5 when you go from $-\infty$ to zero on the horizontal axis.

Now the area between values of -1 and 0 is the same as that between 0 and 1.

The area covered at 1 is 0.84.

The area covered at 0 is 0.5.

Hence the area between 0 and 1 is 0.34.

Standard deviations and normal distributions

Table 7.2 Area under the normal curve

The area given is: $\int_{-\infty}^{z} \frac{1}{\sqrt{(2\pi)}} e^{-\frac{1}{2}z^2} dz$

	0.00	0.01	0.02	0.03	0.04	0.05	0.06	0.07	0.08	0.09
0.0	0.5000	5040	5080	5120	5160	5199	5239	5279	5319	5359
0.1	0.5398	5438	5478	5517	5557	5596	5636	5675	5714	5753
0.2	0.5793	5832	5871	5910	5948	5987	6026	6064	6103	6141
0.3	0.6179	6217	6255	6293	6331	6368	6406	6443	6480	6517
0.4	0.6554	6591	6628	6664	6700	6736	6772	6808	6844	6879
0.5	0.6915	6950	6985	7019	7054	7088	7123	7157	7190	7224
0.6	0.7257	7291	7324	7357	7389	7422	7454	7486	7517	7549
0.7	0.7580	7611	7642	7673	7704	7734	7764	7794	7823	7852
0.8	0.7881	7910	7939	7967	7995	8023	8051	8078	8106	8133
0.9	0.8159	8186	8212	8238	8264	8289	8315	8340	8365	8389
1.0	0.8413	8438	8461	8485	8508	8531	8554	8577	8599	8621
1.1	0.8643	8665	8686	8708	8729	8749	8770	8790	8810	8830
1.2	0.8849	8869	8888	8907	8925	8944	8962	8980	8997	9015
1.3	0.9032	9049	9066	9082	9099	9115	9131	9147	9162	9117
1.4	0.9192	9207	9222	9236	9251	9265	9279	9292	9306	9319
1.5	0.9332	9345	9357	9370	9382	9394	9405	9418	9429	9441
1.6	0.9452	9463	9474	9484	9495	9505	9515	9525	9535	9545
1.7	0.9554	9564	9573	9582	9591	9599	9608	9616	9625	9633
1.8	0.9641	9649	9656	9664	9671	9678	9686	9693	9699	9706
1.9	0.9713	9719	9726	9732	9738	9744	9750	9756	9761	9767
2.0	0.9772	9778	9783	9788	9793	9798	9803	9808	9812	9817
2.1	0.9821	9826	9830	9834	9838	9842	9846	9850	9854	9857
2.2	0.9861	9864	9868	9871	9875	9878	9881	9884	9887	9890
2.3	0.9893	9896	9898	9901	9904	9906	9909	9911	9913	9916
2.4	0.9918	9920	9922	9925	9927	9929	9931	9932	9934	9936
2.5	0.99379	99396	99413	99430	99446	99461	99477	99492	99506	99520
2.6	0.99534	99547	99560	99573	99585	99598	99609	99621	99632	99643
2.7	0.99653	99664	99674	99683	99693	99702	99711	99720	99728	99736
2.8	0.99744	99752	99760	99767	99774	99781	99788	99795	99801	99807
2.9	0.99813	99819	99825	99831	99836	99841	99846	99851	99856	99861
3.0	0.99865	99869	99874	99878	99882	99886	99889	99893	99897	99900
3.1	0.99903	99906	99910	99913	99916	99918	99921	99924	99926	99929
3.2	0.99931	99934	99936	99938	99940	99942	99944	99946	99948	99950
3.3	0.99952	99953	99955	99957	99958	99960	99961	99962	99964	99965
3.4	0.99966	99968	99969	99970	99971	99972	99973	99974	99975	99976
3.5	0.99977	99978	99978	99979	99980	99981	99981	99982	99983	99983
3.6	0.99984	99985	99985	99986	99986	99987	99987	99988	99988	99989
3.7	0.99989	99990	99990	99990	99991	99991	99992	99992	99992	99992
3.8	0.99993	99993	99993	99994	99994	99994	99994	99995	99995	99995
3.9	0.99995	99995	99996	99996	99996	99996	99996	99996	99997	99997

Hence the area between −1 and 0 is also 0.34.

Therefore the probability of a data point falling between the mean and minus one standard in a normally distributed data set is 0.34.

This brings us to a second key observation of normally distributed data sets. That is as follows:

The probability of data falling between plus and minus one standard deviation of the mean in a normal distribution is 0.68.

Finally you might like to establish for yourself the fact that **the probability of data falling within ±2 σ of μ in a normal distribution is 0.95.**

Using the area under the normal curve

The normal curve has been shown to approximate to both symmetrical **and** non-symmetrical binomial distributions when n is large. Hence the area under the normal curve can be used to avoid the use of the general binomial equation to obtain probabilities of particular outcomes.

However it should be recalled that the horizontal axis under the normal curve is plotted as:

$$X = \frac{x - \mu}{\sigma}$$

and it is usually worth recalling that for a binomial probability function the mean μ and standard deviation are given by:

$$\mu = np \text{ and } \sigma = \sqrt{(npq)}.$$

For example, if the overall probability of people responding to corporate credit card mailshots is one in four what is the probability of getting between 70 and 80 responses when a department of 300 employees is mailed with relevant information?

If we had the binomial distribution for $(1/4 + 3/4)^{300}$ plotted as a histogram the probability we require would be the area represented between 69.5 and 80.5 on the horizontal axis.

However, if we use the normal curve to approximate this we could use the area between

Standard deviations and normal distributions

$$\frac{69.5 - \mu}{\sigma} \text{ and } \frac{80.5 - \mu}{\sigma}$$

now $\mu = np = 300 \times \frac{1}{4} = 75$ (for the binomial distribution).

and $\sigma = \sqrt{(npq)} = \sqrt{300 \times \frac{1}{4} \times \frac{3}{4}} = 7.5$

$\therefore \dfrac{69.5 - 75}{7.5} = -0.73$ (using these values on the normal curve).

and $\dfrac{80.5 - 75}{7.5} = +0.73$

Now from Table 7.2 the area under the normal curve from minus infinity to +0.73 is 0.7673 and minus infinity to zero is 0.5 Hence from zero to +0.73 the area is 0.2673 and by symmetry the area from −0.73 to +0.73 is 2 × 0.2673 = 0.5346.

Therefore the probability of getting between 70 and 80 responses to the mailshot is 0.5346.

What is the probability of getting more than 85 responses?

Well we need the difference between the area at plus infinity down to

$$\frac{85 - 75}{7.5} = 1.33$$

The area at +1.33 is read from Table 7.2 as 0.9082.

The area at plus infinity is 1 and so the area above 1.33 is 1 − 0.9082 = 0.0918.

Hence the probability of more than 85 responses is less than 9.2 per cent.

Fluctuations in steady data and dealing with proportions

A typical project problem is to see if a fluctuation in steady data is significant.

Seasonal fluctuations are usually obvious. You don't need statistics to explain increases in turkey sales before Christmas and Thanksgiving!

138 Standard deviations and normal distributions

However what about the problem of seeing if last month's sales increase was due to the advertising campaign. If it is, you ought to repeat it as soon as possible. If it is just a chance fluctuation then maybe you should hold back till next month.

What you need to calculate is the deviation of the proportion of sales that month and compare it to deviations that might occur by chance.

For example if you have been selling 740 000 units a year at very steady monthly figures and your monthly sales immediately rise to 63 000 after a sales campaign, is this significant?

Well you expect to sell $\frac{740\,000}{12}$ units per month = 61 666 units per month.

Hence the numerical deviation for the month in question is
63 000 − 61 666.
This is 1334 units. Quite considerable but is it significant?

Null hypothesis

In this type of problem it is common to set up a hypothesis that helps to keep the logic under control. It is usual to set up a hypothesis that says there is **no difference**
i.e. a null result.

Here the hypothesis would be that 'the advertising campaign made no difference to the sales figures and that the slight improvement occurred by chance'.

Converting this data into the relevant proportions.

Expected proportion of sales per month = $\frac{1}{12}$ = 0.0833

Actual proportion during the month in question = $\frac{63\,000}{740\,000}$ = 0.0851

The deviation in the proportion from the expected figure = 0.0851 − 0.0833 = 0.0018.

Now the sales proportion for any particular month is just a sample that will vary. If it varies, it has a mean and therefore a standard deviation. The distribution could be compared to a normal distribution.

Standard deviations and normal distributions

Now for a binomial distribution the mean of the outcomes x is given by $\mu = \Sigma Px = np$ (see earlier) where P is the probability of any particular outcome, p is the intrinsic probability of success and n is the number of possible outcomes. For outcomes expressed as proportions x can be replaced by

$$\left(\frac{x}{n}\right)$$

$$\therefore \mu_p = \Sigma P \frac{x}{n} = \frac{1}{n}\Sigma Px \quad \text{(since } n \text{ is a constant)}$$

substituting from above

$$\mu = \frac{1}{n} \times np$$

$$= p$$

$$\therefore \mu = p$$

i.e. the mean of a distribution of proportions is the intrinsic probability that establishes the original distribution.

Similarly the standard deviation for original values is given by

$$\sigma_x^2 = npq = \Sigma P(x-\mu)^2 = \Sigma P(x-np)^2$$

so for a proportion $\left(\frac{x}{n}\right)$ replaces x and the mean $\mu = p$.

$$\therefore \sigma_p^2 = \Sigma P\left(\frac{x}{n} - p\right)^2$$

(where p represents the mean of the distribution of proportions).

$$= \Sigma P\left(\frac{1}{n}(x-np)\right)^2$$

$$= \frac{1}{n^2}\Sigma P(x-np)^2 \quad \text{(since } n \text{ is a constant for the original outcomes } x\text{)}$$

but $\Sigma P(x-np)^2 = npq$ (where x is the original outcome)

$$\therefore \sigma_p^2 = \frac{1}{n^2} npq$$

$$\therefore \sigma_p = \sqrt{\frac{pq}{n}}$$

i.e. the standard deviation of the distribution of proportions is related in this way to the intrinsic probabilities and the number of possible outcomes in the distribution of original outcomes.

Now the probability (p) of a sale in the month in question is simply 1/12 and the probability (q) of the sale being in any other month is (11/12).

Therefore the standard deviation of the proportions σ_p is

$$\sigma_p = \sqrt{\frac{\frac{1}{12} \times \frac{11}{12}}{740\,000}} = \sqrt{\frac{0.0833 \times 0.9167}{740\,000}}$$

$$= 0.00032$$

But the deviation of the proportion from the mean for the month in question has already been found to be 0.0018. Hence the critical ratio of these deviations is

$$\frac{0.0018}{0.00032} = +5.6$$

Now, as can be seen from Table 7.2 the probability of deviations of more than $+3\,\sigma$ from the mean is less than 0.002 in a normal distribution. Hence a deviation of this magnitude is very unlikely to have occurred by chance. Hence you should rebook the advertising immediately. That is if the discounted cash flow for the improvement is worthwhile (see Chapter 2).

Formally you would state that the Null Hypothesis has not been upheld.

Levels of significance in surveys

A very common survey result is to state that x per cent of the people surveyed would prefer a particular outcome, given the opportunity. There are two weaknesses in this statement. First it is only arrived at by asking a proportion of the population and second it will be part of a distribution of possible answers that will have their own distribution about a mean value.

If you assume that this distribution follows the normal curve you can use Table 7.2 to put ranges on the spread of values expected. Assuming that the distribution is 'two-tailed' these values are:

5 per cent of the distribution lies outside the range − 1.96 to + 1.96
2 per cent of the distribution lies outside the range −2.32 to + 2.32
1 per cent of the distribution lies outside the range −2.57 to + 2.57
0.1 per cent of the distribution lies outside the range −3.29 to +3.29.

For example in a survey of 2 000 people 40 per cent favoured a particular washing powder. Between what limits would you report the data if you wanted to have (i) a 95 per cent chance of being correct (ii) a 99 per cent chance of being correct?

Now the survey has a distribution of answers given as proportions with a standard deviation given by

$$\sigma_p = \sqrt{\frac{pq}{n}}$$

Now we really need to know the intrinsic probability that a person chosen at random will prefer the powder. However if we assume that $p = 0.4$ and $q = 0.6$ for the 2000 people (n) to be representative of the whole population then:

$$\sigma_p = \sqrt{\frac{0.6 \times 0.4}{2000}} = 0.01095$$

Now for a distribution of the proportions

$$\frac{\text{Actual deviation}}{\text{Standard deviation}} = \text{value on horizontal axis of the normal distribution at the appropriate significance level}$$

For 95 per cent significance, (5 per cent outside the range) using the figures above

$$\frac{\text{deviation}}{0.01095} = \pm 1.96$$

∴ deviation = ±0.022 (rounding up)

Hence the range = 0.4 ±0.022

As a percentage the range equals = 37.8 per cent – 42.2 per cent

Hence we could be 95 per cent sure that the actual proportion of results lie in this range.

For 99 per cent significance (1 per cent outside the range),

$$\frac{\text{deviation}}{0.01095} = \pm 2.57$$

∴ deviation = ± 0.029 (rounding up)

Hence range = 0.4 ± 0.029

As a percentage the range is 37.1% – 42.9 per cent.

Hence we could be 99 per cent confident that the actual proportion of people who prefer the washing powder is in this range.

Standard error of a mean

A further statistical result which will be of considerable use in a later section relates to the fact that the mean of a sample cannot be expected to be equal to the mean of the population it is drawn from. There are many ways of drawing smaller samples from a larger population and the means of these samples will have their own distribution.

The standard deviation (σ_m)(also known as the standard error) of the distribution of the means of simple samples of size n taken from a population is given by

$$\sigma_m = \frac{\sigma}{\sqrt{n}}$$

where σ is the standard deviation of the population. Notice how it depends on the standard deviation of the population and the size of the sample taken. In project work the major use of this concept is in a long term project control methodology and it is used to set warning limits that are brought into play when samples are taken from continuous production lines or service operations for quality control purposes.

The significance of correlations between factors in projects

So far we have considered the significance and accuracy with which we can make observations on single values or sequential values of the same intrinsic information. The statistical complications have come about from the uncertainty surrounding the estimate and the distribution of data around a central most frequent value.

Another key area in project work is to assess the influence of one variable within the project upon another. This is the typical 'cause and effect' in commonly used terms.

When we suspect that two variables may be linked, it is common practice to graph out a **scatter diagram** between them.

Figure 7.5 shows a graph of heights against weight for freshers in a UK university engineering department. As expected there is a general trend with taller people tending to weigh more. This is formally stated as a correlation between height and weight.

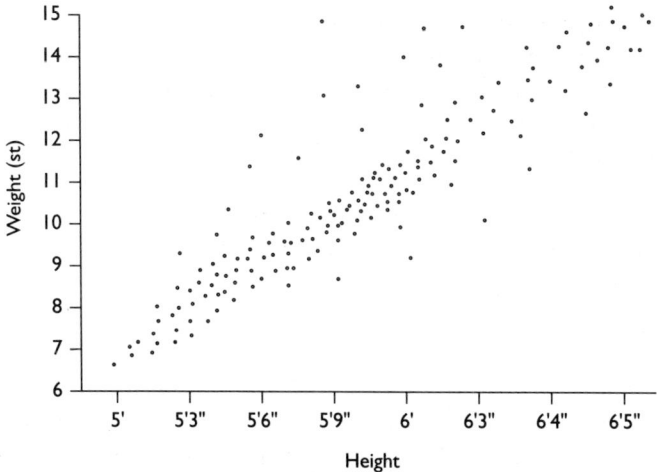

Fig. 7.5 Scatter diagram of height versus weight for university freshers

Different levels of correlation reveal themselves in scatter diagrams. In Fig. 7.6 four comparisons are made with low correlation, medium correlation and high correlation between x and y going from left to right.

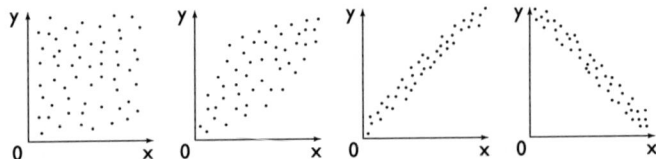

Fig. 7.6 Low, medium and two high correlations between variables x and y

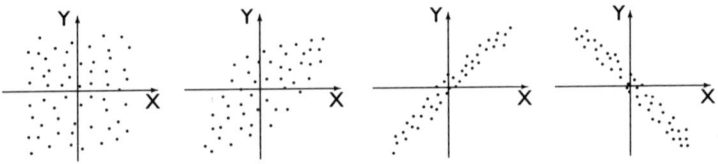

Fig. 7.7 Identical data to Fig. 7.6 but with each value reduced by its mean value

The data is more convenient if it is central around the origin for both variables and so Fig. 7.7 is a plot of the same data but plotted as X versus Y where all the values have been reduced by their mean value.

Hence $X = x - \bar{x}_a$

$Y = y - \bar{y}_a$

the bar and 'subscript a' symbols mean that the arithmetic means of all the x values and all y values have been employed.

Clearly the two correlations on the right-hand side show the strongest links. One is shown with a positive correlation and the other with a negative correlation since the Y values fall with X.

The summation of the products 'XY' reflect the link between the two values for positive and negative correlations and so ΣXY is related to the correlation coefficient r. However this coefficient should not increase simply due to the sample size and so the average is taken.

Second, as a coefficient, it should not have units. Hence dividing

the product by the standard deviation of the x values, σ_x and the standard deviation of the y values σ_y achieves this aim.
Hence

$$r = \frac{\frac{1}{n}\Sigma(x - \bar{x}_a)(y - \bar{y}_a)}{\sigma_x \sigma_y}$$

r can take values in the range

$$-1 \leq r \leq +1$$

and the closer to 1 (in either sense) then the stronger the correlation.

Example

Shown in Table 7.3 are the sales figures (x) in £M and the hospitality budgets (y) in £k for different regions (A–E) of a large company. What is the correlation coefficient?

Table 7.3

Region		A	B	C	D	E
Sales £M	(x)	1	4	5	6	9
Hospitality budget £k	(y)	2	5	6	8	9

To begin with we need $\bar{x}_a$ and $\bar{y}_a$, then we shall need σ_x and σ_y

$\Sigma x = 25$, $n = 5$. $\therefore \bar{x}_a = 5$.

$\Sigma x = 30$, $n = 5$. $\therefore \bar{y}_a = 6$.

x	y	$x - \bar{x}_a$	$y - \bar{y}_a$	$(x - \bar{x}_a)^2$	$(y - \bar{y}_a)^2$	$(x - \bar{x}_a)(y - \bar{y}_a)$
1	2	−4	−4	16	16	16
4	5	−1	−1	1	1	1
5	6	0	0	0	0	0
6	8	1	2	1	4	2
9	9	4	3	16	9	12
$\Sigma = 25$	$\Sigma = 30$			$\Sigma = 34$	$\Sigma = 30$	$\Sigma = 31$

$$\sigma_x = \sqrt{\frac{\sum(x - \bar{x}_a)^2}{n}} = \sqrt{\frac{34}{5}}$$

$$\sigma_y = \sqrt{\frac{\sum(y - \bar{y}_a)^2}{n}} = \sqrt{\frac{30}{5}}$$

Noticing that the different units used cancel out.

$$\therefore r = \frac{\frac{1}{n}\sum(x - \bar{x}_a)(y - \bar{y}_a)}{\sigma_x \times \sigma_y} = \frac{\frac{31}{5}}{\sqrt{\frac{34}{5}}\sqrt{\frac{30}{5}}} = 0.97$$

Hence there is a very good correlation between the annual sales and the hospitality budget! As with all statistical evidence, the interpretation needs expertise. Clearly there may be a causal link between the hospitality budget and sales performance. The hospitality bills could be part of a well-run sales and marketing plan. However it is not the role of the statistician to investigate the underlying principle but to just report the findings to those who are interested.

Defining the correlation and other statistical comparisons

Notice that the correlation scatter diagrams have not had lines drawn on them. This would imply that the form of the relationship is known. However it is beyond the scope of this book to develop these statistical tests and the reader is pointed in the direction of linear and multiple regression analyses available within standard commercial software and covered in depth in many texts devoted solely to statistics.

Similarly, the tests of significance based on the normal distribution can be developed further in two areas that are relevant to project work. They are tests that can determine if a sample is significantly different to the population it was taken from and tests to see if one sample of data distributed in a normal fashion is signifcantly different to other samples that have normally distributed data.

This text is now completed by using the statistics that have been introduced in order to examine process control techniques that can be applied to the long-term implementation of project recommendations.

Statistical process control applied to projects

The principles of statistical process control have their origins in continuous mass production. Here the major concern is to monitor the key aspects of the production process and make adjustments **during** the production process to ensure that the production output remains within pre-determined limits.

This approach is fundamentally different from the long established method of first setting up the process, producing a large batch and then rejecting a certain portion after an inspection procedure. Craftsmen have always made periodic adjustments to their equipment and procedures. However the advent of microprocessor control systems has enabled mathematically based automatic control systems to be used in a continuous fashion to make many minor adjustments without stopping the production process.

Simultaneously, computerized systems within the service industries enable the data needed to quantify many service-based procedures to be collected and analysed.

For example, as fast food orders are keyed into the 'till' it is possible to log the time. If the time is also logged when the bill is paid then the distribution of service times for restaurants, tables and individual employees could be monitored. Clearly this has staff motivation advantages and gives the management team 'hard data' to back-up both rewards and disciplinary procedures. However it is still difficult to programme courtesy and enthusiasm into the system! It is not usually advantageous to be fast but rude. At a higher level of data analysis sophistication it might be possible to examine if certain orders showed strong correlation to unexpectedly long service times. This could generate an investigation to determine why.

At the time of writing it is well-known that supermarket check-out personnel and anyone who uses a telephone system to process enquiries and make sales can have their work rates and success rates monitored.

However the major advantage of statistical process control (SPC) is to accept that there will be minor variations in performance. As usual these variations will be distributed about a mean value. SPC offers a methodology to monitor this data in a continuous fashion and to immediately compare it to warning values. It is then possible to assess if the variations are due to the acceptable distribution within data or evidence of the process approaching or exceeding the pre-determined limits. In such a case the associated closed-loop control can initiate adjustment procedures. In an engineering production line this might mean re-setting control devices. In the service sector it might mean opening up another check-out point or sales station.

Example

It might be a claim of an airport authority, in a new promotions drive, that 'We will get your plane to its stand and the exit doors opened 9 minutes ±3 minutes after it has touched the tarmac.

First of all lets look at the factors that might affect this claim which is typical of the service sector. These are ranked in terms of their potential influence on variability.

(1) Communication and co-ordination within the airport.

(2) Communication between air traffic control and captain.

(3) Level of traffic.

(4) Weather conditions.

(5) Size of aircraft.

(6) Skill of ground crew.

(7) Skill of air crew.

As is usual, the outcome of the co-ordinated sequence of events necessary to park an aircraft and get the doors safely open should approximate to a normal distribution. Here time is the variable. However there are two minor points that are obvious limitations to the approximation.

Standard deviations and normal distributions 149

(i) The lower limit is physically fixed and corresponds to the situation when the aircraft is taxied at the maximum speed allowed, lands at the most convenient runway location, is allocated to the closest stand and all other events work perfectly.

(ii) There is no physical upper limit to the time taken.

Hence we should expect the distribution to be slightly skewed to the high side of the distribution and the mean value to be less than the mid-range value. Let's call the variable the TAT (taxiing arrival time).

The first topic to address is the system/process capability. This attempts to quantify the inherent mean value of the major variable and the inherent standard deviation as a measure of variability.

If you had access to a large amount of past data concerning landing times and docking times at your airport or a comparable airport and the system had not changed significantly, this could be used to get the mean TAT for the 'population' and its standard deviation.

However this is unlikely to be the case and you would therefore need an alternative approach. One method is to take small samples under a variety of operational conditions.

Shown in Table 7.4 are samples of 10 TATs under light, medium, heavy traffic and under night-flying and poor weather conditions. It is not essential to identify each set but just to ensure that you have covered the variety of operational conditions during the sample acquisition. As can be seen nine samples of ten TATs have been taken and the arithmetic mean $\bar{x}$ and the standard deviation **of each sample** have been calculated.Because of the small sample sizes the standard deviation has been calculated using the alternative formula given.With large samples the difference between the formulae is insignificant.

Now it is assumed that the samples are each normally distributed and that variations in the times within each sample reflect the variability inherent within the system itself. The conditions prevailing during each sample are assumed to be constant and result in the changes observed in the mean values. The standard deviations are for each sample and are calculated around the appropriate mean value. Hence the standard deviations recorded for each sample are now averaged to give an overall inherent system variability.

From Table 7.4 it can be seen that the mean TAT varies considerably. However, for each set of readings, it is a reasonable assump-

150 Standard deviations and normal distributions

tion that externally imposed conditions, such as traffic levels, weather, etc. are constant. Hence the standard deviations which are calculated against the different mean values reflect the intrinsic variability of the taxiing and docking process. By calculating

$$\sigma_i = \sqrt{\frac{\Sigma \sigma^2}{n}}$$

this value can be found

$$\sigma_i = \sqrt{\frac{1.591^2 + 1.301^2 \ldots\ldots\ldots\ldots 1.252^2}{9}}$$

$$\therefore \sigma_i = 1.15 \text{ minutes}$$

The tolerances of ±3 minutes therefore lie $3.0/1.15 = 2.61$ standard deviations away from the mean value due to the intrinsic variations within the process alone.

Table 7.4

TAT mins	1	2	3	4	5	6	7	8	9	10	$\bar{x}$	σ
SAMPLE												
1	8.1	7.6	11.1	7.5	9.1	9.5	9.5	8	12	11	9.34	1.591
2	8.8	9.9	12.2	9.1	11	10.5	9.9	8.2	8.3	8.6	9.65	1.301
3	7.3	7.5	7.7	7.9	8.5	7.5	7.9	9.1	8.8	8.1	8.03	0.596
4	9.9	9	10.5	9.9	12.1	11.1	9.7	10.6	11.3	9.9	10.40	0.909
5	12.5	11.6	11.2	6.9	8.8	8.6	9.4	11.1	10.3	9.2	9.96	1.685
6	6.6	7.3	7.2	6.6	7.3	7.8	7.1	6.9	8.1	6.7	7.16	0.499
7	8.9	9.4	9.5	10.4	11.1	10	8.3	11.1	10.6	12	10.13	1.129
8	8.3	7.9	9.1	8.8	8.4	8.3	9.9	9.6	7.7	8	8.60	0.835
9	1.2	11.2	11.4	9.5	9.4	9.2	7.7	10.4	10.1	10.4	10.13	1.252

Therefore from Table 7.2 the probability of a TAT being outside the upper of these limits is 0.005. Hence from an operational view, the probability of a TAT being outside the claimed limit due to the intrinsic variability, is just over 0.5 per cent (1 in 200) represented by the single-tail area under the normal distribution beyond 2.61 standard deviations from the mean.

The next problem is that the arithmetic means of the samples vary and the overall arithmetic mean of the data gathered may be calculated to be 9.27 minutes.

Clearly the means of the samples are distributed about a central value. We should be concerned on how this distribution compares to our perceived mean of the population i.e. 9 minutes.

As you know the standard deviation of the distribution of the means is also known as the standard error and is given by

$$\sigma_c = \frac{\sigma}{\sqrt{n}}$$

Where σ is the standard deviation of the population and n is the sample size. Hence for these samples the standard error is

$$\frac{1.15}{\sqrt{10}} = 0.364$$

This estimate can be used to monitor the process. If during operation samples are periodically taken for 10 consecutive landings their mean values can be plotted on a control chart. Figure 7.8 shows the control chart for the data in Table 7.4.

Just as before we only expect values to fall more than three standard errors away from the target mean very rarely. Hence it has been possible to draw control lines at upper and lower limits of

$$\pm 3 \times 0.364 \cong \pm 1.1 \text{ minutes}$$

As can be seen from the samples, all TATs are currently well within the claim of 9 mins ±3 mins. However, there should be some concern that three sample mean values out of nine are on or above the upper control limit.

Recalling from Table 7.2 that the probability of this happening as an individual event is less than 2 in 1000 it seems likely that this data is not fully described by this approach.

There are two distinct possibilities for this

(i) The mean of the population is not 9 but closer to the overall value of 9.27 recorded in these samples.

(ii) The data and samples do not approximate to a normal distribution and the system is inherently more variable than expected.

As can be seen from Fig. 7.8 increasing the mean TAT to 9.27 minutes would include the three upper values within the upper control limit. However two values would still be outside the control lines on the minimum side. If the process were under control and the data belonged to the same overall distribution that could be approximated by a normal distribution then values should be outside both these control lines once in five hundred. Hence the data probably needs careful scrutiny and reassessment.

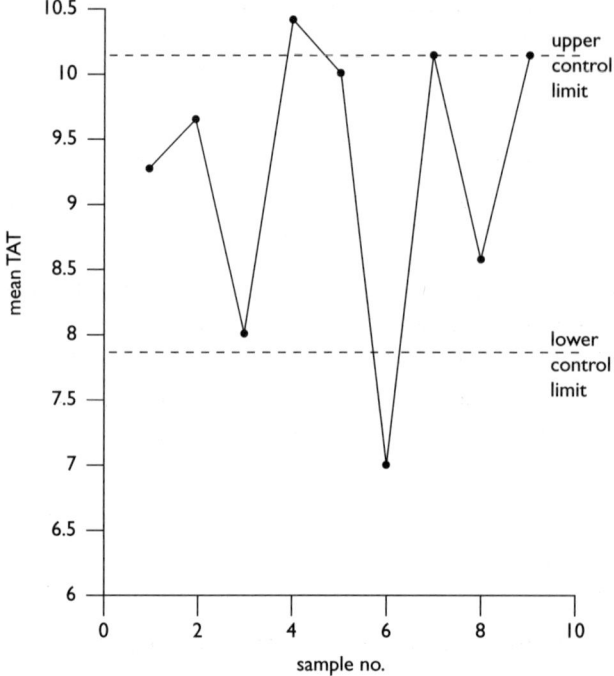

Figure 7.8 Control chart for mean TATs

It seems clear that the TATs do not belong to the same population and that there may be two distinct groups: those with TATs over 9 minutes and those with TATs below 9 minutes. This is where statistics should cease and an inspection of the experiment should begin. It might reveal that two terminals were involved with significantly different intrinsic TATs or one group of landings relied on mobile disembarkation steps and the other on automatic gates attached to the terminals or the precise turn-off point from the main runway was crucial or even the wind direction that

Standard deviations and normal distributions

causes planes to land in different directions caused two distinct distributions of the original data.

In the absence of experimental evidence it is recommended that you now reassess the data by treating it as two separate groups. Better still get down to your local airport and collect some of your own. However get permission first as the local security will be alerted by binoculars and stop watches collecting data about aircraft movements. This is part of the hazards of project work!

Objectives

After reading this chapter you should be able to do the following.

1. Calculate the standard deviation from a set of grouped or ungrouped data.
2. Describe how a set of data can be approximated by a normal distribution and calculate the probability of a particular value occurring from a given table of areas under the normal curve.
3. Set up a Null Hypothesis and test it for the significance of fluctuations in steady data, collected as absolute values or proportions
4. Add ranges to survey data for particular levels of significance.
5. Show how the intrinsic deviations in a process or service can be established.
6. Estimate control limits for process or service control purposes.

Index

Acceptance sampling, 96
Advertising, 37
Advertising, campaign costs, 38
Advertising, NPV in, 39
Arithmetic mean, 84

Bar chart, 22, 72

Cash flow, 30, 47
Correlation, 76, 143
Correlation coefficient, 144
Cost, minimum, 52
Costing, average gross annual rate of return, 32
Costing, average net annual rate of return, 33
Costing, discounted cash flow, 34
Costing, net present value, 35
Costing, payback period, 31
Critical path analysis, 44
Customer satisfaction, 56

Delay event, 46
Dependent variable, 75

Direct costs, 59
Discounted cash flow, 34
Dummy event, 44

Equations, of lines on graphs, 82
Exponential law, 83

Factorial notation, 112
Fixed costs, 9, 59
Fluctuations in data, 137
Franchise, 58
Function, aesthetic, 8
Function, cost, 6, 9, 19
Function, primary, 8
Function, secondary, 8
Function, tree, 7, 8
Function, triangular comparison, 18
Functional importance, 6, 7

Geometric mean, 87
Graphs, 80
Graph, gradient of, 80
Graph, intercept on axis, 80
Graph, origin, 80

Index

Graph, planar, 77, 79
Grouped data, 88

Histogram, 22, 72

Idea, evaluation action chart, 22, 26
Idea, generation, 20
Idea, stimulator prompt list, 21
Independent variable, 75
Indirect costs, 59

Just–in–time production, 48, 58

Mailshots, 136
Mean of probability distribution, 127
Median, 86
Mid-range value, 86
Minima by differentiation, 52
Mode, 87

Net present value, 39
Node, 41
Node slack, 41
Null Hypothesis, 138

Pareto analysis, 23, 27
Pie charts, 69
Power Law, 83
Prestige potential, 13
Probability, definition of, 93
Probability, distribution, 107
Probability, problems in, 117

Probability, rectangular distribution, 108
Probability, rules, 97, 99
Probability, symmetrical binomial distribution, 109
Project, risks, 104
Project, selection, 10

Quality Function Deployment, 2

Redundancy, 64
Response rates, 110, 136
Risk analysis, 102

SPC, 147
Scatter diagram, 143
Shift working, 61
Significance, 143
Spot check, 97
Standard deviation, 123
Standard deviation of probability distribution, 128
Standard error, 142
Statistical process control, 117, 147
Sub contracting, 62, 63

Tabulated data, 67

Unit costs, 60

Value, 1, 6
Value, analysis, 1

Value, engineering, 1
Value, prestige, 7
Value, service, 7
Variable costs, 9, 59
Variables on graphs, 75, 77

Weighted average, 89

X factor, 11, 13